THIS IS
YOUR BRAIN
ON
PARASITES

THIS IS
YOUR BRAIN
ON
PARASITES

How Tiny Creatures Manipulate
Our Behavior and Shape Society

KATHLEEN McAULIFFE

An Eamon Dolan Book
Houghton Mifflin Harcourt
BOSTON   NEW YORK

www.hmhco.com

Library of Congress Cataloging-in-Publication Data
Names: McAuliffe, Kathleen, author.
Title: This is your brain on parasites : how tiny creatures manipulate
our behavior and shape society / Kathleen McAuliffe.
Description: Boston : Houghton Mifflin Harcourt, 2016. |
"An Eamon Dolan book."
Identifiers: LCCN 2016002949 (print) | LCCN 2016009925 (ebook) |
ISBN 9780544192225 (hardback) | ISBN 9780544193222 (ebook)
Subjects: LCSH: Nervous system—Diseases. | Parasitology. | Microbiology. |
BISAC: SCIENCE / Life Sciences / Biology / Microbiology. | PSYCHOLOGY /
Psychopathology / Schizophrenia. | MEDICAL / Microbiology.
Classification: LCC RC346 .M36 2016 (print) | LCC RC346 (ebook) | DDC
612.8—dc23
LC record available at http://lccn.loc.gov/2016002949

Book design by Rachel Newborn

Printed in the United States of America
DOC 10 9 8 7 6 5 4 3 2 1

To my family, and in loving memory
of my sister Sharon McAuliffe, a very talented
science writer who died way too young

CONTENTS

THIS IS
YOUR BRAIN
ON
PARASITES

INTRODUCTION

WE LIKE TO THINK of ourselves as in the driver's seat, choosing where to go, whether to speed up or slow down, when to switch lanes. We make the decisions and bear the consequences. This is a convenient, even necessary belief. If we jettison the notion of free will, the laws that hold people accountable for their actions begin to crumble. The world becomes an unruly or even terrifying place. Alien beings that turn us into zombies, bloodthirsty vampires, and sex-crazed robots are standard sci-fi fare precisely because they evoke the horror of losing control or, worse, becoming slaves to creatures bent on exploiting us for their own gain. So it's disconcerting to think that an invisible passenger might also have a hand on the steering wheel, vying to move us in one direction when we'd rather go another. When we let up on the accelerator, an unseen foot presses harder.

Parasites are like that invisible passenger. Adept at outwitting our immune systems, they sneak aboard our bodies and then the devilry begins. They cause rashes, lesions, aches, and pain. They eat us from the inside out; use us to incubate their young; sap our energy; blind, poison, maim, and sometimes kill us. But that's not the full extent of their clout. Some parasites have another trick up their sleeves — an awesome hidden power that astounds and confounds even scientists who study them for a living. Simply stated, these parasites are masters of mind control. Whether as tiny as a virus or as big as a six-foot-long tapeworm, they have found all kinds of devious methods to manipulate the behavior of their hosts, and that includes, many researchers now strongly suspect, humans.

The impetus for this book was a discovery on the Internet. I'm a science journalist and one day while foraging for interesting topics to write about I stumbled across information about a single-celled parasite that targets the brains of rats. By tinkering with the rodent's neural circuits — exactly how is still a matter of fervid study — the invader transforms the animal's deep innate fear of cats into an attraction, thus luring it straight into the jaws of its chief predator. This is a felicitous outcome not only for the cat but also, I was stunned to learn, for the parasite. It turns out the feline gut is exactly where the organism needs to be to complete the next stage of its reproductive cycle.

This revelation got me thinking about my own cat, who was fond of dropping dead rodents at my feet. Horrified as I was by this habit, I could not help admiring her hunting prowess. Now I wondered if it was she who was so clever or the parasite.

As I continued reading, more surprising news greeted me: The microscopic organism is a common inhabitant of the human brain because cats can transmit it to us when we come in contact with their feces. Perhaps the parasite was meddling with our brains too, speculated a Stanford neuroscientist associated with the research. I contacted him to find out what he meant and was pointed in the direction of a biologist in Czechoslovakia. "He's a bit of a wild man," he warned me, "but I think it would be worth your while to speak to him." I called Prague and over the span of an hour was told a tale as bizarre as any I've heard in my profession. It occurred to me on several occasions that the person at the other end of the line might be a kook, but I pushed those thoughts aside and kept listening because it was impossible not to. I'm a sucker for a great story and this one had all the elements of a first-rate medical thriller. It was by turns creepy, scary, weird, and inspiring. What's more, if true, it had important health ramifications.

After the conversation ended, I called around to other experts on this cat parasite for a reality check. I did this rather sheepishly at first, out of fear of sounding gullible. But one source after another said that the Czech's ideas, though far from proven, deserved serious scrutiny. His human studies — and the odyssey that led him down that path of inquiry — became the basis of a lengthy article I wrote for *The Atlantic* and are described in a chapter here, along with his most up-to-date results, so you can draw your own conclusions. (A word of caution: Before you get to that section, please do not panic and give away a pet cat. As I will explain in more detail, there are much more effective ways to protect against the infection than parting with a cherished companion.)

Over the course of investigating the topic, I came across many other stories of parasitic mind control; I learned of parasites that force their hosts to be their personal bodyguards, babysitters, chauffeurs, servants, and more. Sometimes scientists understand how they accomplish these feats; other times, they're left scratching their heads. It seemed to me that neurosurgeons and psychopharmacologists could learn a lot from parasites.

Once I became aware of their antics, it was hard to look at the world outside my window in the same way again. Behind the scenes of the spectacle we call natural selection, I was surprised to learn, parasites are often directing the action, influencing the outcome of the battle between predator and prey. Insights into their stagecraft gave me a radically different perspective on ecology, evolutionary biology, and the spread of mosquito-borne scourges like malaria and dengue hemorrhagic fever.

While parasites' coercive tactics have many disturbing implications for humans, the news from this front is not all bleak. Some microbes may actually improve our mental health. And invaders with sinister aims will have to contend with much more than our immune systems.

Mounting research suggests that hosts have developed powerful psychological defenses against parasites. Scientists call this mental shield the behavioral immune system. Experiments show that it kicks into action in situations where the threat of infection is high, prompting the organism in peril to respond in prescribed ways to reduce its risk. A simple example is a dog that reacts to being hurt by licking its wound, thus coating the injury with saliva rich in bacteria-killing compounds. In smart primates like humans, however, it appears that our behavioral defenses have become tied to increasingly abstract and symbolic ways of thinking. Many habits and traits that seem far removed from pathogens — such as our political beliefs, sexual attitudes, or intolerance toward people who break societal taboos — may arise at least in part from a subconscious desire to avoid contagion. There is even evidence that the presence or absence of germs in our immediate surroundings — indicated by such signs as a rancid odor or filthy living conditions — can influence our personalities.

Directly or indirectly, parasites manipulate how we think, feel, and act. In fact, our interaction with them may shape not only the contours of our minds, but also the characteristics of entire societies, perhaps explaining some puzzling cultural differences between parts of the world where pathogens are an omnipresent threat and areas that have dramatically lowered that risk through vaccination programs and improved sanitation. Numerous lines of evidence suggest that the prevalence of parasites in our broader communities influences the foods we eat, our religious practices, whom we choose as mates, and the governments that rule us.

The science behind these claims is still young. Some findings are preliminary and may not hold up to scrutiny. But the research is massing quickly and the outlines of a new discipline are clearly taking shape. This newly emerging field has been christened *neuroparasitology*. But don't be deceived by the label. While neuroscientists and parasitologists currently dominate this endeavor, it is increasingly drawing in investigators from fields as diverse as psychology, immunology, anthropology, religious studies, and political science.

If pathogens' impact on our lives is really so far-reaching, why has it taken us so long to discover this? One likely reason is that, until recently, scientists underestimated the sophistication of parasites. Over most of the past century, the complicated life cycles of these organisms, coupled with their puny size and concealment inside the body, made them exceedingly difficult to study. Largely out of researchers' ignorance, parasites were presumed to be backward, degenerative life forms. Their inability to survive as independent, free-living creatures was seized as proof of their primitive status. The very notion that hosts high up the evolutionary ladder might be jerked around like marionettes by such simpletons — many lacking even a nervous system — seemed absurd.

Until the tail end of the twentieth century, our behavioral defenses against parasites were also assumed to be rudimentary. Indeed, the subtlest of these adaptations — manifested as automatic thoughts and feelings — were overlooked almost entirely, probably because they occur at the periphery of our awareness. Scientists are no more cognizant of subconscious impulses than the rest of us, so this subterranean realm appears to have gone uncharted simply because no one thought to look for it.

Even today, the intimacy and intricacy of parasite-host relationships take many neuroscientists and psychologists by surprise. Laymen are often dumbfounded by how nature could have given rise to parasitic manipulations in the first place; some stratagems seem so clever and cunning that only a human or an omniscient god could have dreamed them up. The emergence of the behavioral immune system in parallel with such manipulations only adds to the challenge of comprehending the origins of these interactions. So before moving ahead, let's stop to ponder how evolution took this turn.

Parasites and hosts have been competing with each other for billions of years. The first bacteria were parasitized by the first viruses. When larger, multicellular life forms emerged, these microbes in turn colonized them. Meanwhile, parasites continued evolving into a menagerie of distinct forms — roundworms, slugs, mites, leeches, lice, and the like. As life grew in size and complexity, natural selection favored parasites that were the best at evading hosts' defenses, and hosts with the greatest skill in repelling the invaders.

Today, virtually every aspect of the human body's design bears witness to this age-old struggle. Our most visible defense is our skin, which provides a thick barrier to the hordes of microbes that populate its surface. Entry points are especially fiercely guarded: Eyes are bathed in tears that flush out intruders. Ears are lined with hairs to keep out bugs. The nose has a filtration system for screening pathogens out of the air. Invaders that make further inroads will only encounter stiffer resistance. The respiratory tract, for example, produces mucus that traps encroachers. As for any microbes that we swallow in our food, they'll likely meet a fiery death in the cauldron of the stomach, whose industrial-strength acid could literally burn a hole in your shoe. Should all these defenses be breached, immune cells will rush into the battle. This army is led by sentries that flag the intruder, and they're followed by white cells that devour it and still other cells that record the enemy's markings so that new regimens can swiftly be called up should the body encounter the same foe again.

With firepower like that you'd think humans would always be on the winning side. But parasites have huge advantages over us. Their population size dwarfs our own by staggering numbers, and their rapid replication rates ensure that there will always be a lucky few with mutations that will give them the upper hand. The battle between hosts and parasites is an unending arms race.

In this intensely competitive environment, any parasites that by chance hit on ways to modify the behavior of a host so as to enhance their own transmission — perhaps, for example, by nudging it a wee bit closer to the parasites' next host — would very swiftly multiply. Since hosts can't evolve as quickly to thwart every new trick parasites deploy against them, their best chance for survival is to acquire traits that offer them broader protection. Mutations that prompt an animal to feel repelled by common sources of contagion — for instance, murky green water, a dung heap, or other members of its flock acting strangely — might serve that function. The beauty of such psychological adaptations is that they shield against not one, but hundreds or even thousands of infectious agents. That's a lot of bang for the buck — an opportunity that evolution is unlikely to have passed up. In humans, moreover, instinctual responses that protect against infection would also be amplified and embellished through learning and cultural transmission, further leveraging their benefit. It's a good bet that's exactly what happened.

Though lions, bears, sharks, and weapon-wielding humans may populate our nightmares, parasites have always been our worst enemy. In medieval times, one-third of Europe's population was decimated by the bubonic plague. Within a few centuries of Columbus's arrival in the New World, 95 percent of the indigenous population of the Americas had been wiped out by smallpox, measles, influenza, and other germs brought in by European invaders and colonists. More people died in the 1918 Spanish flu epidemic than were killed in the trenches of World War I. Malaria, presently among the most deadly infectious agents on the planet, is arguably the greatest mass murderer of all time. Experts estimate the disease has killed half of all people who have roamed the planet since the Stone Age. New insights into how parasites spread among us and the hidden power of our minds in countering this tsunami-size threat could yield huge benefits.

One is that it might suggest innovative ways to block the dissemination of much-dreaded infectious agents. Another hope is that discoveries in neuroparasitology will expand our knowledge of the root causes of mental disturbances that we don't normally associate with parasites, possibly leading to advances in their prevention and treatment. The discipline's greatest promise for the near future, however, is its capacity to enrich our understanding of ourselves and our place in nature. Certainly, findings from this frontier raise provocative questions: If pathogens can fiddle with our minds, what does that say about our responsibility for our own actions? Are we really the freethinkers we imagine ourselves to be? To what extent do parasites define our identity? How do they affect moral values and cultural norms? In the final chapter of this book, I will attempt to salvage the concept of free will. But be warned: it will take quite a beating in the meantime.

1

Before Parasites Were Cool

I T'S NOT EASY BEING a parasite. Sure, you get a free meal. But the life of a moocher still comes with plenty of stresses. You have to be able to adapt to the environment inside one, two, or, if you belong to a class of parasitic worms known as trematodes, three different hosts — habitats that can be as different from each other as the Earth is from the moon. And getting from one to the next can be a logistical nightmare. Imagine you're a trematode that spends part of life inside an ant but can only sexually reproduce inside the bile duct of a sheep. Ants aren't on a sheep's normal menu, so how do you make it to your next destination?

The answer to that question is what set Janice Moore on her life's path. In 1971, she was a senior at Rice University in Houston sitting in an introductory course on parasitology taught by a titan in the field, Clark Read, a lanky man with a commanding presence and an odd style of lecturing. He would puff away on a cigarette and seemingly free-associate, drawing students into his passion with fascinating details about different species of parasites that he presented with no discernible regard for logic or order. But he was a gifted storyteller who could evoke the lives of parasites so richly that you could almost picture what it was like to be one. He also knew how to spin a good mystery, which was how he ensnared Moore.

She couldn't imagine how to get an ant into a sheep's mouth in spite of Read's admonishment to "think like a trematode!" In fact, no one could, because the solution the parasite lit upon is absurdly improbable: It invades a region of the ant's brain that controls its locomotion and mouthparts. During the day, the infected insect behaves no differently than any other ant. But at night, it does not return to its colony; instead, it climbs to the top of a blade of grass and clamps onto it with its mandibles. There, it dangles in the air, waiting for a grazing sheep to come by and eat it. If that doesn't happen by the next morning, however, it returns to its colony.

Why doesn't it just stay attached to the leaf? asked Read, scanning the classroom as if he expected his students to discern the trematode's logic. Because otherwise, he told his rapt audience, the ant will fry to death in the noonday sun — an undesirable outcome for the parasite, which will perish with it. So up and down the ant goes, night after night, until an unsuspecting sheep eats the ant-laden blade of grass, and the parasite finally ends up in the sheep's belly.

Read's tale stunned Moore. The trematode called to mind a comic-book arch villain who controls minds with a joystick, causing law-abiding citizens to rob banks and commit other crimes so the villain can take over the world. The report of the trematode's astonishing feat came from a German study done in the 1950s, but, thrilling Moore, Read had just learned of research being done on a different organism that was producing findings similar to the Germans'.

The protagonist of this tale was a thorny-headed worm — a parasite with a spiky head and a flaccid body that looks like a five- to ten-millimeter worm-shaped sac. Before assuming its adult form, the parasite must mature inside tiny shrimplike crustaceans that live in ponds or lakes and that usually burrow into mud at the first sign of trouble. For the next stage of the worm's development, however, it needs to get inside the gut of a mallard, beaver, or muskrat — all creatures that live on the water's surface and feed on the crustaceans. To determine how the stowaway manages to jump ship, John Holmes, a former student of Read's who had become a professor at the University of Alberta, and his graduate student William Bethel brought crustaceans into the lab. Infected ones, they discovered, did exactly what they shouldn't. Instead of diving downward when agitated, they shot to the surface and skittered around, all but crying, *Look at me!* If that failed to draw attention, they clung to vegetation that waterfowl and aquatic mammals liked to eat. Some, Moore was amazed to learn, even attached themselves to the webbed feet of ducks and were promptly swallowed.

Another intriguing detail grabbed her attention. Occasionally, the Canadian investigators found, the crustaceans harbored a different species of thorny-headed worm. When infected with this variety, their tests showed, the crustaceans also swam upward in response to any disturbance, but they congregated in well-lit areas frequented by scaup (deep-diving ducks) — as it turned out, that particular parasite's next host.

Many interactions between predators and prey, thought Moore, were not what they appeared to be but rather were "rigged" by parasites. Perhaps biologists, who couldn't see what was happening out of view, had been hoodwinked! What's more, if parasites were not just swinging a sledgehammer, directly killing and sickening hosts, but also bringing ill upon them by subtly changing their behavior, the ecological implications were enormous. It meant that these tiny organisms were taking animals out of one habitat and putting them in another, with unknown effects that would ripple through the food chain.

When the class ended, she rushed up to Read. "This is what I want to study," she announced, brimming with excitement. He applauded her decision as an adventurous one and they hatched a plan for her future. You'll need to get a master's in animal behavior and then you should get a PhD in parasitology, he advised, and she did exactly that.

Four decades later, she looked back on that day with amusement. "I was bright-eyed, enthusiastic, and totally ignorant of the obstacles in the way," she said, breaking into a deep-throated laugh at the thought of her youthful optimism. Vivacious, with short wavy hair, Moore still has a trace of a Texas twang and she has a vibrant, confident style. Now a professor of biology at Colorado State University, she has arguably worked harder than anyone else to awaken the biology community to the game-changing nature of parasitic manipulations and encourage a new generation of scientists to take up that cause. Her pioneering studies — and, more important, her writings — have shone a spotlight on the myriad ways parasites bend hosts to their will and on their subversive, often underappreciated role in ecology. Predators, in her view, may not always be the supreme hunters nature documentaries suggest they are. A significant portion of their catch of the day may be low-hanging fruit brought within their reach courtesy of parasites. Why, after all, work hard for a meal when it will come to you? Perhaps the most heretical notion of the field she helped found is simply that one should not assume animals are always acting of their own volition. Numerous crustaceans, mollusks, fish, and "literally truckloads of insects," according to Moore, "are behaving weirdly because of parasites." Mammals like ourselves appear to be less common victims of their manipulations, but that belief may derive from ignorance, she cautioned. This much she's certain of: An undiscovered universe of animal behavior will yet be traced to parasites. Their meddling, in her view, is just harder to prove in some species than others.

Moore and a growing cadre of like-minded scientists are making progress in their mission, but it's been a long haul — as the reason for our first meeting in the spring of 2012 underscored. We'd both traveled thousands of miles to a bucolic corner of Tuscany, Italy, to attend the first-ever scientific conference devoted solely to parasitic manipulations. Sponsored by the prestigious *Journal of Experimental Biology*, the historic event drew a few dozen researchers from all over the world — a tribute to how far the discipline had come but also an occasion to reflect on how much further it would have to go to attain a stature commensurate with its importance. While Moore was delighted that their work was starting to make waves beyond their tiny specialty, she was frustrated that many scientists still failed to grasp how pervasive parasitic manipulations were in nature. Even in many quarters of biology "they're often viewed as little more than cute tricks or one-of-a-kind novelties," she complained.

Another challenge facing neuroparasitology is semantic. Defining what exactly constitutes a manipulation, she said, can be tricky in itself. Technically, she and most of her colleagues concur, the term refers to a behavior a parasite induces in its host that benefits the parasite's transmission at the expense of the host's reproductive success. But that seemingly straightforward definition can be surprisingly murky when applied to the real world. If a cold germ makes you cough uncontrollably, for example, is that your body trying to clear the infection from your lungs or the parasite tickling the back of your throat so that you'll spread the germ? Or consider this: Barnyard hens are probably more inclined to eat crickets infected with parasites that damage the insects' muscles, since those crickets are slower and thus easier to catch. The parasite needs to get into the hen to reproduce, but is it truly manipulating the cricket or merely hurting it? By contrast, few people on hearing of the ant that climbs a grass blade in response to a trematode invading its brain would discount the insect's behavior as a mere side effect of illness. So how far do you expand the definition of *manipulation*?

Moore admits it's not always an easy call. But it amazes her that even when the evidence for a behavior being a manipulation is clear cut, you might not know it from the cautionary tone of many researchers' reports. After one scientist's talk, she observed, "Almost every paper I've reviewed in the last year has the same disclaimer, almost verbatim: 'The alteration in the host's behavior may be due to a manipulation by the parasite or a pathology.' When are we going to have the confidence to say something is not just a byproduct of disease but obviously a manipulation?" Her colleagues nodded in agreement.

Afterward, I asked her why researchers might be timid about expressing their views. "Because reviewers almost always insist that you stick in that qualifier" or they won't accept it for publication, she replied. Ideas that challenge the status quo tend to be resisted, and "pathology," she said, "is the default explanation" — the conservative fallback position, even if it's the least likely possibility.

The rigid either/or thinking that traditional-minded biologists often bring to the topic also rankles Moore. The conduct of parasites and hosts locked in battle cannot always be "neatly sorted into piles," she said. Maybe your cough represents both your body's effort to expel the germ *and* the parasite's determination to spread itself. Even enemies can share the same goals. Insisting that a parasite-induced behavior perfectly fit the profile of a manipulation to warrant scientific interest is equal folly, in her view. To illustrate her point, Moore noted that one of her graduate students had recently found that dung beetles dug shallower burrows and ate 25 percent less dung when they became infected with roundworms. "That's of huge ecological importance," she emphasized. "Australia actually had to import dung beetles because they were up to their ears in dung. Here's a beetle that's an ecosystem engineer that is itself being engineered by the parasite. So we submit this to the *Journal of Behavioral Ecology* and the editor does not even send it out for review. He writes back, 'This is obviously simply a case of pathology' — as if, in the context, it even mattered. It was exasperating!"

If Moore sometimes sounds piqued about preaching to the unenlightened, it's understandable. Especially at the outset of her career, she often felt like a lone wolf howling in the wilderness. Her ideas were not so much disparaged as ignored. At the time she had her epiphany in Clark Read's class, many biologists turned their noses up at parasites, deeming them too primitive and repugnant to be worthy of examination. Birds with fancy plumage and majestic mammals like elephants and lions were considered more appropriate subjects of study. Parasites — insofar as they received any attention — were almost exclusively the domain of veterinarians or medical researchers seeking to stem the tide of epidemics like malaria and cholera. Few people were concerned about their ecological impact, much less the possibility that they could boss around more estimable animals.

Into this world stepped Moore, a young woman espousing that very view. She was not only a maverick but also — by her own admission — "hopelessly naive."

After getting a master's in animal behavior at the University of Texas at Austin, she proceeded to Johns Hopkins University in Baltimore to begin her PhD in parasitology, at which point she assumed she could dive straight into her area of interest. "I hadn't a clue how research was actually conducted — that graduate students don't get to set their own research agenda but rather are expected to work on whatever is the pet interest of their advisor." As it turned out, that person wanted her to throw her energy into studying the biochemistry of tapeworms — a topic that held no appeal for her. Compounding her difficulty adjusting to Hopkins, Moore was the only female graduate student in the department and felt isolated from her peers. As a result, she had little sense of what others perceived to be important issues in the field, which ironically may have helped as well as hindered her development as a scientist. When I asked Moore if her ignorance might have freed her to think outside the box, she shot back: "I didn't even know there was a box!"

She was a misfit in still other ways. Science is inherently reductive; its ethos is to break big problems down into smaller chunks that can be more easily attacked. But Moore has always been a big-picture person. She sees connections between almost everything she learns and likes to synthesize information. As an undergraduate she agonized over choosing a major, ultimately settling on biology owing to its breadth. The study of every living thing on earth shouldn't constrain her too much, she thought. When it came time to decide on a specialty within that field, parasitology and animal behavior attracted her for similar reasons. "This just seemed like about the most stuff you could ever pull together, and I was in a stage of life where I hadn't a clue that it's also extremely difficult to pull things together, which is why they tend not to be together," she said, again bursting into laughter at her younger self's gung-ho, climb-any-mountain mentality.

Her grand vision of manipulative parasites reordering food chains was thrilling to entertain, but she had no notion of how to design an experiment to test the sprawling concepts that crowded her mind. Hopkins, which had strong parasitology and ecology departments, had initially seemed the perfect place to learn that skill. But to Moore's disappointment, the groups were not closely connected. "They perceived themselves to be studying very different things," she explained. Because she had no guidance as to how to bridge those disciplines, her goal of placing parasitic manipulations in a broader context seemed far beyond her reach.

Further frustrating her, whenever she attempted to open others' eyes to the possibility that parasites might be puppeteers, she was not warmly received. At a seminar about the ecology of marine snails in the intertidal zone, she asked the scientist giving the lecture if he'd looked to see if trematodes were in the mollusks. Infected snails tended to be found in different places than those free of the parasite, she explained, citing a paper she'd just read. The researcher became visibly upset. From his perspective, he already had his hands full charting numerous factors impinging on the behavior of the snail — migrating predators, shifting currents, daily fluctuations in temperatures, and more — and here she was suggesting that he should attend to yet something else. Moore was not unsympathetic to his view — studying parasites in the field remains a daunting task to this day — but at the time, his reaction came as a heavy blow.

Unable to see a way forward, Moore decided to drop out of Hopkins at the end of her first year. The Christmas prior, while back in Texas, she'd made plans to connect up with Read, her former professor, who'd already indicated he might be willing to let her study parasitic manipulators under his supervision. But shortly before they were supposed to meet, he died unexpectedly of a heart attack, leaving Moore saddened and academically adrift. She made numerous inquiries at other universities in search of a PhD program that might offer her a comparable opportunity, but neuroparasitology was not even a gleam in a scientist's eye at that time. Even John Holmes, the Canadian scientist whose lab showed that some crustaceans were acting at the behest of parasites, was not actively pursuing that line of study. It was a side interest, he explained. Moore had hit a dead end.

With no good options, she took a job at the University of Washington as a lab technician for an entomologist whose interests did not overlap with her own. But her luck was about to change. The scientist, Lynn Riddiford, was a rarity of that era, a woman who had risen to the top of her profession, and she proved to be a great role model. At her side, Moore learned how research projects were conceived, funded, and carried out — essentially, the nuts and bolts of being a successful scientist. She came away from the experience empowered and with new confidence in her ideas. Perhaps because she took herself more seriously, other people did too. After a three-year detour, Moore was offered a place at the University of New Mexico in a unique PhD program that provided funding for students to design their own research projects.

It was a big opportunity and she didn't want to blow it. By then she knew that she wouldn't be able to connect all the dots, that it would be a triumph simply to identify any parasitic manipulation not yet recognized, especially if she could show those manipulations made the hosts more appealing to their predators under field conditions. From Riddiford, she'd also learned the importance of designing a tight experiment, ideally one with a simple premise that was easy to execute. After searching through academic papers and textbooks for most of a semester, she finally believed she'd found the ideal subjects for her study. The parasite was a type of thorny-headed worm that cycled between two exceedingly common and easy-to-observe hosts, starlings and pillbugs (or roly-polies, as children often call them, owing to their tendency to curl up into balls when touched). With only a few hunches to go by, Moore theorized that the parasite would make the pillbug behave in ways that would increase the insect's likelihood of being eaten by a starling.

Her experimental apparatus consisted of a glass pie plate with nylon mesh stretched across most of the top and an inverted pie plate for its cover. She placed a mixture of infected and uninfected pillbugs on top of the mesh and then introduced a different type of salt on each side of the divider, creating one chamber with low humidity and one with high. Pillbugs that harbored the parasite, she discovered, were much more likely to gravitate toward the low-humidity zone. In the wild, dry areas coincide with exposed locations, so she assumed the behavior of the infected pillbugs would make them more vulnerable to predation. In another experiment, she built a shelter by placing a tile on top of four stones, one at each corner. Infected pillbugs preferred to be out in the open more than uninfected ones did. In yet another experiment, she covered one half of a pie plate with black gravel and the other with white gravel to test whether the parasite affected the host's ability to camouflage itself. Because pillbugs are black, she theorized that infected ones would be more likely to hang out on the white gravel, where they'd be conspicuous to birds. Indeed, that's exactly what she found.

She'd proven her thesis in the lab, but would her findings hold up in the field? Owing to the difficulty of studying parasites in their natural habitat, no scientist had been able to measure the ecological impact of a manipulation. But Moore had a clever plan to do just that. She set up nest boxes for starlings on the campus during breeding season. She tied pipe cleaners around the throats of starling nestlings, just tight enough to prevent them from swallowing but loose enough not to hurt them. Then she collected the prey that their parents fed them and proceeded to dissect any pillbugs among the day's capture. She found that almost one-third of starling babies had been fed infected pillbugs, even though less than 0.5 percent of pillbugs in the vicinity of the nest boxes harbored the parasite. Clearly, the changes the parasite induced in the habits of its hosts had made them far more attractive prey.

One or two examples of parasites with remarkable manipulative powers can easily be brushed aside as bizarre aberrations — intriguing, to be sure, but hardly more than a footnote to our understanding of natural selection. But more examples begin to look like a trend. When Moore's results appeared in the journal *Ecology* in 1983, they attracted attention not only for that reason but also because of a broader development sweeping biology. After long being shunned as disgusting low-lifes, parasites were starting to be embraced as objects of fascination and even admiration. As Moore put it, "They became cool."

Just why this happened isn't clear — science, like all fields, is subject to fads — but coinciding with her starling studies, a flurry of papers pointing out the ecological importance of parasites began appearing in scientific journals, and they were penned by giants in evolutionary biology like Robert May, Roy Anderson, and Peter Price. Around the same time, another prominent evolutionary biologist, Richard Dawkins, published a popular book, *The Extended Phenotype,* that touched more closely on the theme of parasitic manipulation. In the book he argued that whether a gene is passed down depends not just on how it affects the characteristics, or phenotype, of the body in which it resides but also on its impact on other animals. In that category, he cited as one example natural selection favoring parasites that change a host's behavior to propagate their own genes.

Parasites' sudden rise in popularity worked in Moore's favor. Editors at *Scientific American,* a magazine with a reputation for presenting cutting-edge research, invited her to write an overview article that would place her pillbug findings in a larger framework. In addition to highlighting the German and Canadian studies, she combed the scientific literature for other noteworthy cases of parasitic manipulations that had been ignored or overlooked and then explained their significance in lively, accessible prose.

"One of the most familiar literary devices in science fiction is alien parasites that invade a human host, forcing him to do their bidding as they multiply and spread to other hapless earthlings," she began the article in the May 1984 issue of the publication. "Yet the notion that a parasite can alter the behavior of another organism is not mere fiction. The phenomenon is not even rare. One need only look in a lake, field or a forest to find it."

Soon the manipulation hypothesis, as it became known, was being discussed with great interest in scientific circles. As Louis Pasteur famously remarked, "Chance favors the prepared mind." Once word spread that parasites might be dictators in disguise, more people began to notice animals behaving strangely, and inquiring minds wondered whether infectious organisms might be to blame.

Despite scientists' excitement about the idea, however, the field's popularity proved fleeting. The practical challenges of doing the research quickly dimmed enthusiasm for it. Observing animal behavior is an arduous undertaking even before parasites enter the picture. It can entail spending endless hours submerged underwater in scuba gear, hanging from a harness at the top of a forest canopy, or searching through swampland with a flashlight in the middle of the night. Since a parasite may have two or three hosts, just working out the details of its life cycle can be a Herculean task. To add to the challenge, estimating infection rates in each population usually means catching dozens or hundreds of potential hosts and either drawing blood from the animals, collecting their feces, or killing and dissecting them. Then — assuming you clear all those hurdles — comes the really hard part: determining whether the hitchhiker actually is manipulating the host and, if so, how and for what purpose. That's best done in the lab, but many animals aren't inclined to go about their everyday activities in captivity. Humans may be more cooperative subjects, but scientists who suspect that mental illness or other aberrant behavior might be linked to parasites hit an even bigger obstacle: they can't infect people with their bug of choice and then watch to see if the subjects' habits or dispositions change.

Not surprisingly, researchers with the patience and perseverance to pursue such work are hard to find, which is why even today there's a tendency to focus on the interplay of predator and prey and ignore the hidden passenger that might have a very different agenda than the vehicle it's riding inside. Nonetheless, by the turn of the millennium, scientists had succeeded in uncovering several dozen parasitic manipulations affecting hosts across virtually every branch of the animal kingdom. Always the synthesizer, Moore in 2002 compiled all known cases into a book, *Parasites and the Behavior of Animals,* still viewed as a bible in the field. Her goal in writing it was to inspire creative thinking about how parasites work their black magic and to uncover unifying principles. How often, she tried to determine, do they target the central nervous system of the host? Do closely related species employ similar coercive strategies? Could very complex manipulations have simple underpinnings? Above all, her ruminations focused on a question that had captivated her since she was a student in Clark Read's class: Can you predict the behavior of animals by the parasites inside them?

Moore is still trying to answer those questions. Patterns are emerging, but details remain sketchy, she admits. And the task at hand keeps getting bigger. Hundreds more parasites are now suspected of being manipulators, and the true number, she speculates, may be in the thousands. "We just haven't bumbled across them," she said. And not just because of the difficulty of studying animal behavior or the taboo against experimenting on humans. Perhaps our biggest handicap is that we're imprisoned by our senses. Quite simply, we rely too heavily on our eyes for our understanding of the world. In the talk Moore gave at the conference, she emphasized that point by recounting the story of the discovery of bat echolocation.

Since the eighteenth century, investigators have known that blindfolded bats can deftly navigate between silk wires but bats whose ears are sealed crash to the ground. Yet for over a hundred and fifty years, scientists refused to believe that the animals could hear what humans could not. In the early 1940s, advances in detecting sounds in the ultrasonic range demonstrated that bats could hear the echo of their own cries. Yet the notion that they might use that skill to navigate was still not fully accepted until the military declassified documents after World War II revealing the development of radar and sonar.

With that in mind, said Moore, it's worth noting that most of the manipulations we know of today can be viewed with the naked eye. The intermediate host positions itself against a high-contrast background, or moves around frenetically, or is someplace where it wouldn't ordinarily go. Because this attracts human attention, we can readily understand why it would be noticed by a predator that is the parasite's next host. But what if a parasite effectively puts a target on an animal's back by altering aspects of its behavior that are invisible to our senses? Perhaps, for example, it induces its host to lay down a scent trail that our noses can't smell, or causes it to emit sounds beyond our hearing range, or prompts it to expose body parts that look drably colored to us but are vivid to the parasite's next host. That marked animal inviting predation may even be one of us. As we shall later see, some parasites that cause terrible scourges are now suspected of enhancing their transmission by altering human body odor. "Given these possibilities," said Moore, "how can we fail to wonder what manipulations we are missing in this wild world of information that lies just beyond our senses?"

A colleague who stepped up to the podium immediately after her, biologist Robert Poulin from the University of Otago in New Zealand, agreed that scientists were missing thousands of manipulations but, interestingly, presented a reason different from any she'd proposed. Many manipulators, he pointed out, may cause merely the tiniest shifts in a host's normal habits — something that could be overlooked when scientists compare the average behavior of a host population to the uninfected group. For example, parasites may slightly modify how often animals go one place or another, alter the time of day they're most active, or prompt hosts to act in an ordinary manner but in the wrong context — an infected bird pecks the ground while the rest of the flock takes wing, for instance. "Predators are highly attuned to anything that makes prey stand out in the least way" so this strategy would likely be highly effective, he suggested. Also, this kind of minor tweak should not be a difficult one to pull off, hence evolution might have favored this simple gambit. What this implies for humans is that we may need to study behavior with a finer-tooth comb to detect parasites' influence on us — for example, attempting to link suspected meddlers not just to flagrant mental illness but also to more subtle shifts in personality and habits that are well within human norms.

Fortunately, it's now easier to test these kinds of theories and get answers to some of the big questions posed by the science. As the discovery of bat echolocation demonstrates, it often takes technological advances before new frontiers can be broached, and, heartening in that regard, science is finally starting to catch up with the sophistication of parasites. Over the past decade, the tools for getting at the mechanisms behind manipulations have advanced dramatically. As a result, researchers have far better methods for imaging parasites within the host's body and for identifying the genes, neurotransmitters, hormones, and immune cells involved in these behavioral changes. There are few manipulations that have been worked out in their entirety, but, as the next few chapters will reveal, scientists now have some excellent clues. That's great news, for if we are to "think like a trematode," we will need to understand its tricks.

2

Hitching a Ride

THE ANECDOTAL REPORTS WERE WILD. Crickets that normally inhabited the forest floor and didn't swim were leaping headfirst into ponds and streams. Frédéric Thomas suspected that a worm seen wriggling out of the insect's body as it drowned was behind the cricket's suicidal impulse, but the only way to be sure was to go to New Zealand, where the phenomenon had been reported. In 1996, Thomas, an evolutionary biologist at the University of Montpellier in France, applied to the French government for money to investigate the matter, confident his proposal would be funded. Though he'd only just gotten his PhD, he already had fourteen scientific publications to his name — a prodigious number for such a young scientist — so he assumed he'd be a shoo-in for a grant. What's more, an animal acting so flagrantly against its own best interests was clearly a topic worthy of study — or so he thought. Stunning him, the Centre National de la Recherche Scientifique (CNRS) — France's equivalent of the United States' National Science Foundation — turned down the proposal. He was so angered by their decision, he told me, that he decided to go on a hunger strike.

For a moment I thought he was pulling my leg. But his somber expression suggested he was serious. We were chatting on a veranda during a break in the Tuscan conference on parasitic manipulations where I'd met Moore. Thin, with dark tousled hair, Thomas came across as easygoing, confident in an understated way, and charmingly exuberant about his research. Now I wondered if his enthusiasm for his work bordered on crazy. CNRS is by far the largest funder of research in France, so it's not a good idea to piss off the folks there by threatening them. I scrutinized his face. Was there something I wasn't getting? Could I have misunderstood him?

In fact I had. He did not tell the CNRS that he'd go on a hunger strike if the grant wasn't forthcoming, he clarified; he told the president of France. "I sent a letter directly to Jacques Chirac."

You'd assume a low-level bureaucrat would open it, have a hearty laugh, and toss it in a wastebasket. But, amazingly, the import of his message — if not the letter itself — was passed up the chain of command to high-level officials.

Far from generating laughter, his letter appears to have sent the French government into a panic. Officials from the administration were promptly dispatched to his university, where they pressed the chairman of his department to prevent him from carrying out his threat. If Thomas could not be kept from going on a hunger strike, they intimated to the department head, both scientists would pay dearly in lost grants. Evidently the officials were unsettled by the thought of an emaciated Thomas turning public sentiment against the government. They put so much pressure on Thomas that he finally agreed to withdraw his threat.

Deeply discouraged, the biologist was anxiously debating what to do next when a Swiss billionaire named Luc Hoffmann heard of his plight through another scientist and came to his rescue. Known for his philanthropy and strong interest in biodiversity, Hoffmann offered to pay half the cost of the expedition. With that backing, Thomas was able to secure matching funds from the embassy of New Zealand and other sources, including the French government, which gave him a small sum, glad to be rid of him.

Happy to have his troubles behind him, he went off to New Zealand, where he linked up with a team at the University of Otago led by Robert Poulin, an evolutionary biologist whom he greatly admired and who was his source of information about the cricket. The two men immediately hit it off. A tall man with a soft, melodic voice and an amiable disposition, Poulin grew up in a French-speaking part of Canada, so he and Thomas shared a common language in addition to scientific interests. But their cricket investigation never got off the ground. They ran into the kinds of obstacles that Moore had warned often derail studies of parasitic manipulations: The insect came out of its burrows only at night and tended to hide among low bushes, and its green body perfectly matched its surroundings, making it hard to see even when in plain view. Even though the team scoured the landscape with flashlights night after night, often crawling on hands and knees through low brush, they caught only a handful of crickets infected with the worm — nowhere near the number required to run experiments that would produce meaningful results. After battling the French government and traveling thousands of miles, Thomas was forced to admit defeat.

Not one to squander an opportunity, he switched to another scientific project, but before doing so, he sent a photo to a university colleague of a worm emerging from a cricket — "just to give news of me, 'Look what I'm up to.'" The friend posted the picture in his department's coffee station, where a lab technician happened to see it. A cousin of his in Montpellier cleaned pools for a living, he wrote Thomas, and they were full of the worms.

Thomas was highly skeptical. The New Zealand parasite is just one of three hundred hairworms, as scientists refer to this vast category of threadlike organisms, so he assumed the technician was mistaken. But when he got back to France, he met with the technician's cousin and gave him a jar of alcohol in which to put any worms that he found in pools. Thomas figured he'd never see the fellow again, but a week later the man returned with a jarful of worms.

He had collected them from a pool at a nearby resort, he told

Thomas, who was very curious as to how they'd gotten there. "I convinced my wife to go on a romantic holiday in the area because there was a hotel with a very nice restaurant with foie gras, and a hot-water spa nearby," Thomas told me. A mischievous smile spread across his face as he relayed this, hinting that he might have brought her there under false pretenses. After they had a lovely meal, he went on, he did not retire to the hotel room with her but retrieved test tubes from the trunk of his car and returned to stake out the pool. Soon after, he saw a cricket approach the pool, and his first impulse was to step on it. Given his long journey to this moment, you'd think he'd resist the urge. But indeed he did crush it, right there on the patio floor. Lifting his foot, he observed a three-inch-long worm spilling out of the insect's crumpled body — the very same worm that had parasitized the crickets in New Zealand! He'd traveled halfway around the world to study a parasite and host that could easily be found virtually in his own backyard.

A few minutes later another cricket appeared and lunged into the pool. He leaned over to get a closer view. A "living hair" snaked out of its body. "I thought I'd cry," he said. "It was really incredible. Seventy-five kilometers from my home in Montpellier is probably the best place in the world to study this cricket." At night, against the turquoise backdrop of the illuminated pool, the violent birth of the worm was as easy to see as actors lit up on a stage. In the subsequent field studies that he and his students conducted at an open-air pool near a forest in Avène-les-Bains, they had abundant opportunities to observe the electrifying spectacle. In addition to inhabiting crickets, the worms also turned up inside grasshoppers and katydids that similarly developed a mysterious attraction to water. Indeed, the "enchanted" insects came in droves. On a typical summer's night, over a hundred flocked to the pool.

In an effort to understand how the hairworm choreographed this remarkable show, Thomas's team began research on its life cycle. How could an aquatic worm get into a land-dwelling cricket in the first place?

Once the hairworms break free of their hosts, the team discovered, they mate in the water, and the females then lay a string of eggs, which develop into larvae. As they swim around, they bump into the larger larvae of mosquitoes and hop aboard them, hiding inside them as tiny cysts (think of nested Russian dolls). When those mosquito larvae morph into winged adults, they take flight, carrying the parasite with them to land, where they die and are eaten by crickets. The dormant cyst then springs to life, eventually growing into a worm three or four times the length of the insect's body when uncoiled.

They'd sketched out the broad outline of the hairworm's life cycle, but they rarely caught crickets in the act of jumping into a natural body of water — something they very much wanted to see because ponds and streams, unlike swimming pools, are teeming with fish and frogs. Alerted by the splash, these predators would presumably snap up the flailing insect. The hairworm would have to be very quick in exiting its host's body to avoid being eaten itself. Could the creature be that fast?

To gain more clarity on the issue, Thomas purchased a frog from a medical supplier and used an aquarium in his lab to simulate the natural conditions in which the parasitic manipulation occurred. Then he introduced an infected cricket and stood back to watch the show. In the blink of an eye, the frog ate the cricket — hairworm and all. Thomas was mystified. How did the worm escape predation in the wild when, in the simulation, it died with its host? It made no sense at all.

A little while later, he got his answer. The worm squirmed out of the frog's mouth and swam away! After being swallowed, Thomas discovered, it actually got as far as the frog's stomach before turning around and traveling back up the animal's throat. Or sometimes it escaped through the frog's nostrils. When a fish snapped up the infected cricket, the worm exited through the creature's gills. It was an escape artist unrivaled in nature, the Houdini of the animal world.

"No one had ever seen this anti-predator defense before," said Thomas. Poulin had joined our conversation on the veranda, and he put the parasite's feat this way: "It would be comparable to you having a tapeworm, a lion eats you, and then the worm crawls out through its mouth." The team's results were published in the British journal *Nature,* one of the most competitive and prestigious publications in science. "I spent twenty-five euros" — about thirty-five dollars at the time —"for the frog. That was the total cost of the experiment," boasted Thomas.

The biggest mystery of all — how the parasite draws its host to a watery grave — has been the hardest one to crack. But in recent years, his team has uncovered numerous leads, each one more fascinating than the next. An infected cricket, they found, first begins behaving erratically, boosting the likelihood it will stumble into a pond or stream. But as the worm grows in size, eventually consuming almost all of the cricket's innards, something happens that leads the insect to more actively search for water. Was the parasite making its host thirstier? Thomas wanted to know. Or was it doing something else?

To get at the underlying mechanism, his team extracted the worms from the crickets' bodies before and after they dove into the water. His graduate student David Biron, a molecular biologist who is now at the National Center for Scientific Research and the Université Blaise Pascal in southern France, then applied proteomics — a new technique for identifying proteins made by an organism — for insights into the phenomenon.

The results were eye-opening. The worm was producing a raft of neurochemicals that closely mimicked ones normally found in the cricket. "If we don't speak the same language," explained Thomas, "we can't communicate. So if I'm the worm, I want to talk to you in the same language. Natural selection favors worms" that make molecules the cricket can recognize, facilitating "crosstalk" between them. In this way, the parasite can tell the cricket what it wants the insect to do.

More recently, a team led by Biron made another intriguing discovery. Compared to healthy controls, the stricken insects have higher amounts of a protein involved in sight, possibly altering their visual perception. This revelation prompted the French researchers to explore whether crickets harboring the parasite are attracted to light. Indeed they were, whereas the healthy insects preferred the dark. If you're a cricket that lives in the forest, said Thomas, what in your surroundings is brightest of all at night? An open area filled with water — an excellent reflector of moonlight. By tinkering with the settings of the cricket's visual system, he believes, the worm mesmerizes its host. It's effectively whispering to the insect, "Go toward the light."

In an ironic postscript to his story, Thomas now heads a team at the National Center for Scientific Research in Montpellier and clearly is no longer persona non grata in the eyes of the French government. In 2012, his work on parasitic manipulators and other biological topics earned him the CNRS Silver Medal — one of the nation's highest honors for people who have made an outstanding contribution to science.

IT'S HARD TO IMAGINE that a parasite could induce a human to plunge into water, but there is one that does exactly that. It's called the guinea worm and, impressively, it achieves this feat without producing neurochemicals or going anywhere near the person's brain. In fact, it heads in the opposite direction.

The worm, which is now mostly limited to Sudan, gets into people when they drink stagnant water contaminated with water fleas that carry its larvae. The acid in the human stomach kills the water flea, but not the parasites inside it, which develop into worms that slip through the walls of the intestines and mate inside the abdominal muscles. The males, which are only an inch long, then die and are absorbed by the body. But the female grows and grows, eventually stretching a yard in length (I once had the displeasure of seeing one coiled up in a jar of formaldehyde; it looked like a very long strand of spaghetti).

As the worm develops, it snakes through the body's connective tissue toward a lower extremity — typically a foot or calf. After about a year, the female is pregnant with a bustling brood of larvae. To usher them forth into the world, she migrates up to the surface of the person's skin.

Until this point, the parasite has used various types of subterfuge to remain invisible to the immune system, but now she releases an acid that causes the victim's skin to bubble into a painful blister (the disease is called, not surprisingly, dracunculiasis, Latin for "affliction with little dragons"). If she's lucky, this burning sensation will compel the sufferer to dunk the inflamed limb in the nearest body of water. The moment the tapeworm senses the aqueous environment, she breaks through the person's skin and begins disgorging her young through her mouth. Hundreds of thousands of the larvae are ejected with each convulsion. Over the next few days, whenever she comes in contact with water, she again vomits up babies by the thousands. Once released, they swim around until they find a berth inside a new water flea and then repeat the ghastly cycle that will torment more humans — or sometimes the very same ones (people don't develop immunity to the tapeworm, so they can become reinfected).

Just twenty years ago the guinea tapeworm infected 3.5 million people in twenty nations, but today, owing to education campaigns and simple, inexpensive water-filtration systems, it hovers on the brink of extinction, with fewer than one hundred cases of guinea-worm infection occurring annually. Even in the parasite's last holdout, one of the poorest corners of Africa, it is now blessedly rare to see a person racing toward water on a worm's order.

SOME PARASITES DON'T ALTER only the behavior of their hosts; they transform their appearance as well. An outstanding example is the flatworm *Leucochloridium,* a favorite of parasitologists since the 1930s for reasons that will soon become obvious. The parasite replicates inside a bird's digestive system and gets excreted in its waste, so a snail feasting on bird droppings may accidentally ingest the worm's eggs. Once inside the snail, the eggs hatch and eventually grow into long tubes that take over its brain and invade its eyestalks — the first step in the snail's dramatic makeover. As its eyestalks swell in size, their walls are stretched so thin that it's possible to see the parasite within — and what a dazzling sight it is. The worm is bedecked in colorful, pulsating bands, which are in fact pouches packed full of its rambunctious larvae. ("For my own part I could have snails with *Leucochloridia* before my eyes for hours," wrote one early observer mesmerized by the strobe-light effect of the parasite's dancing stripes.) As the snail morphs, it abandons its nocturnal lifestyle and becomes active during the day — very, very active. Polish biologist Tomasz Wesołowski, an expert on the manipulation, clocked one infected animal moving three feet in fifteen minutes — an Olympian speed for a snail. What's more, an infected snail will leave shaded spots on the ground that shield it from view and unwisely mount vegetation, often positioning itself on a high-up leaf with its psychedelic tentacles on full display.

To a songbird overhead, those plump, pulsating stalks look like caterpillar grubs, enticing it to swoop down and peck on them. The victim of the ruse gets a beak full of tiny parasites that will soon reproduce inside its body. As for the snail, it may not only survive the ordeal but also go on to regenerate its eyestalks. However, this may not be as merciful a turn of events as it seems, for the snail may be only a meal away from being reinfected — and getting its eyes pecked out all over again.

A tapeworm that infects brine shrimp is another talented makeover artist. It turns its normally translucent hosts bright pink — and that's not all. It castrates the animal, extends its lifespan, and, possibly by fooling it into thinking it's time to mate, prods it to seek out other infected shrimp. Indeed, the infected crustaceans — each barely the length of a thumbnail — gather in such dense swarms that the water in those spots can form red clouds more than a meter across. This is very convenient for flamingos, which feed on the shrimp, and for the parasite, because the bird is its final host. And thanks to the tapeworm, the long-legged birds can satisfy their appetite simply by dipping their ladle-shaped beaks into the red seafood broth at their feet. Of course, what goes in one end comes out the other, with infected birds eventually releasing a new generation of the tapeworm's eggs into the water.

Animals traveling in large packs, shoals, or flocks, note Nicolas Rode and Eva Lievens, the French scientists who identified the manipulation, are assumed to do so for their own benefit — for example, to find mates, deter predators from attacking, or enhance foraging strategies. Rode and Lievens think it's time to reexamine that assumption. Perhaps far more often than we realize, parasites might be herding their current hosts straight into the jaws of their next hosts.

WHILE MOST PARASITIC MANIPULATIONS come to light because a host is acting bizarrely, occasionally the discovery process follows a different script: a parasite is found tucked up inside an animal's tissues, and then, often acting on a hunch, an investigator looks more closely at the host's behavior and begins to suspect foul play. Such an intuition was the basis of a finding by Kevin Lafferty, now an ecologist with the U.S. Geological Survey at the University of California at Santa Barbara. Lafferty, who looks as lean and fit as a Marine and much younger than his fifty-plus years, grew up in Southern California, where he spent his youth surfing, snorkeling, and scuba diving. To pay his way through college, he got a job that entailed removing mussels from offshore oil rigs. It was laborious work, but he was enthralled by sea life and loved being outdoors, so in search of a higher-paying gig to pursue his passion, he decided to get a degree in marine biology.

Parasites were not initially a special interest of his — in fact, he gave them little thought at all until, early in his career, he taught a class on dissecting fish, sharks, and many other aquatic organisms. Every time he cut open a tissue or organ, "parasites would fall out," he said. "Many specimens had two, three, four, or more. I started to think we were missing a big piece of the picture in trying to understand ecology and the interactions between organisms at different levels of the food chain."

His interest in the environmental impact of parasites led him to study a ribbonlike fluke that sexually reproduces in the gut of egrets, seagulls, and other birds that frequent estuaries in Southern California. The birds shed the fluke's eggs in their droppings, which are eaten by horn snails along the shore. After further maturation inside the snails, the eggs hatch and are excreted. At high tide, the developing parasites get swept into the water, where they latch onto killifish — the most common prey of the shorebirds — invade their gills, and follow nerve tracts up into their heads.

You can find thousands of the fluke's larvae caked onto the surface of a killifish's brain, said Lafferty. He'd been closely following the literature on parasitic manipulations, so the location of the larvae immediately raised his suspicions that the parasite might be meddling with fish minds. Puzzlingly, however, infected fish looked healthy and vigorous. Nothing about their behavior struck him as strange.

Concerned that he might be overlooking a subtle change, he scooped killifish up in a net and released them into a large aquarium. His undergraduate assistant Kimo Morris was then charged with closely observing them to see if he could tell infected fish from their parasite-free peers. After many hours of scrutiny, Morris began to notice a trend. The infected fish were more likely to dart and shimmy along the surface, often rolling over onto their sides — a behavior that presumably would be conspicuous to predatory birds. How much more often did they do this? Morris tallied his observations of individual fish and arrived at a startling number: four times more often. The difference between the two groups was not so subtle after all.

It seemed logical that shorebirds would be drawn to fish behaving so foolishly, but Lafferty and Morris wanted to be sure that their theory held up in the real world. To test it, they gathered a mixture of infected and uninfected fish into an open-air pen that they set up in shallow estuarine water, with one side of the enclosure abutting the shore. Birds could fly into the penned area or wade in unimpeded from the shore. They came at first one by one, then in larger numbers. Three weeks later Lafferty and Morris dissected the remaining animals. Only a few of the healthy fish had been eaten, but almost all of the infected fish were gone.

Observing natural selection play out in this miniature theater, said Lafferty, was highly instructive. Like most fish, killifish are dark on top and light on the belly. "When they roll over on their sides, you see this bright flash — this silver glint. It's almost like someone shining a rescue mirror in your face. The fish that are infected are every bit as healthy as the uninfected fish. They just swim up to the surface and wave hello to the birds that come down and eat them."

To figure out how the parasite could coax its host into acting so imprudently, he and graduate student Jenny Shaw analyzed the neurochemistry of infected fish. They found that the parasite was disrupting the regulation of serotonin, a neurotransmitter that influences the anxiety level of many animals, including humans (the popular antidepressant Prozac alters the metabolism of serotonin). Following that lead, the scientists conducted an experiment in which they traumatized healthy and infected fish by plucking them out of the water for several seconds at a time. Afterward, the healthy fish showed increased activity in serotonin circuits — a sign that they were under acute stress. In stark contrast, the infected fish had a muted response in those brain circuits. "The more parasites the fish had," said Lafferty, "the less stressed-out the fish. This suggests it's so mellow that it doesn't get anxious in a situation that should make the animal fearful. It's less risk averse, like a fish on Prozac."

Most killifish that live near horn snails in Southern California's estuaries will be infected with the fluke by the time they reach adulthood. If we could actually see the parasite inside its hosts as we walked through these marshy areas, the fluke's numbers would astound us, for the snails, the killifish, and the birds it infects are among the most common denizens of these habitats. And if we stood still for a while, we would see the parasites move from one host to the next, acting en masse like a giant conveyor belt, carrying food from the ground, to the sea, to the air, and back down again in an endless loop.

What would happen if the parasites were removed from this picture? Would there be fewer birds in the sky, more fish in the sea? Lafferty doesn't know, but the change would almost certainly have a domino-like effect on the food chain. In some fragile ecosystems where animals struggle to get by on scarce resources, manipulative parasites might even tip the balance toward the survival or extinction of a species. Lafferty recounted joining Japanese biologists who were studying a type of endangered trout in hopes of increasing its numbers. In the fall, the team noticed, the fish were unusually well nourished; their bellies were packed full of crickets. What had made this source of nutrients suddenly so plentiful? A hairworm closely related to the species Thomas has long been studying was sending droves of the crickets into the water in late summer. If it weren't for the parasite, said Lafferty, it's possible the trout might already be extinct.

PARASITIC MANIPULATIONS MAY PLAY a prominent role in determining human population size as well. Some of the world's worst scourges are transmitted by blood-feeding insects whose behavior may in turn be controlled by microscopic infectious agents. I must confess that I was surprised to learn such pathogens could be manipulative — not because I thought they'd lack the means, but rather because I assumed they'd have no motive. After all, these parasites need only wait for a hungry mosquito or fly to come by and bite their current host. Then they're up and away to new digs. What could be easier than that?

Tragically for our species, however, the one-celled parasite that causes malaria — plasmodium — is not so cavalier about its travel arrangements. This pathogen leaves very few aspects of its dispersal to chance. Accumulating evidence suggests it can regulate a mosquito's lust for blood to maximize its own dissemination. Still more impressive, it may alter human odor to enhance our attractiveness to mosquitoes when the parasite is most infectious.

To understand how plasmodium does this, it helps to be familiar with a mosquito's dining habits. To eat, the insect must pierce a human's thick skin with its proboscis and quickly wiggle it around until it hits a blood vessel. Time is of the essence; if the mosquito takes too long to steal a meal, its target may retaliate and flatten it with a smack. Almost as soon as the insect begins drinking, however, the victim's platelets rush to the site and begin to clump up and plug the leak. Sucking becomes increasingly difficult for the mosquito as this cellular debris clogs its feeding tube. To counter that, it injects an anticoagulant that breaks apart platelets, a move that will keep its dinner flow-ing smoothly a few moments longer. Then, skittish lest it get swatted, it promptly sets off for a new patch of skin.

Should plasmodia be consumed along the way, however, this once-voracious feeder soon loses its appetite. Scientists think they know the reason why: plasmodia must reproduce in the gut of the mosquito before their offspring can be transmitted to a person, so if the insect continues feeding during this period, it risks being squashed with no benefit to the parasite. After ten days, however, the parasite's progeny have reached a more infectious stage in their development. At this point, it's very much in the microbe's interest to crank up the mosquito's appetite, which it does by invading the insect's salivary gland and cutting off its supply of anticoagulant. The result is that the mosquito's proboscis quickly becomes gummed up by platelets whenever it attempts to drink. The frustrated insect can't get its normal fill of blood, forcing it to feed on many more hosts to satisfy its hunger. (Incidentally, the bacterium behind the Black Death also obstructs the feeding ability of infected fleas, so when they jump from rats to humans, the insects bite us more vigorously.)

Plasmodium has still more tricks. Once it invades your circulatory system, it interferes with your ability to make platelets, so blood flows more freely when a mosquito comes to dine. In this way, the flying syringe can extract more infected blood to transmit to other people.

As if all these manipulations weren't sufficient to guarantee the parasite's success, it may employ an even subtler form of sorcery. Plasmodia may bring mosquitoes and people together by exploiting the insects' ability to find us by smell. The hairs sprouting from their antennae have sensors that are very good at detecting the carbon dioxide we exhale, the lactic acid in our sweat, and the ammonia of stinky feet. When plasmodia enter your body, they may amplify your natural odors or coax you to produce new scents enticing to mosquitoes.

This claim is controversial — not all research supports it. But a study of Kenyan schoolchildren certainly lends it credence. The investigators began by drawing the students' blood to see which of them harbored the parasite. They then divided the youngsters into a dozen groups, three children in each. Every triad consisted of a healthy child, a second youth at an early, nontransmissible stage of the illness, and a third whose disease had progressed to the infectious stage. Mosquitoes were released into a central chamber connected by pipes to three small tents; inside of each, a single child slept (all of the children were protected from being bitten). Twice as many mosquitoes were drawn to the youngster with transmissible malaria than to those in the other categories. Most intriguingly, when all the kids who had the parasite were given drugs that cleared their infection, the mosquitoes no longer showed a preference for one group over another.

Plasmodia may not be alone in exploiting such tactics. The germ responsible for leishmaniasis, a largely tropical disease that can cause nasty skin sores and damage internal organs, may also make infected people smell more attractive to its insect vector — in this instance, sandflies. In studies of infected hamsters, the microbe changed the composition of the aromatic compounds that gave the animals — and, by extension, perhaps also humans — their distinctive scent.

Interestingly, the mosquito-borne virus that causes the excruciating joint pain of dengue fever (also aptly called breakbone fever) seems to take the opposite tack. Instead of coaxing humans to produce alluring scents, it boosts the insect's ability to track us down. It does this, as best as scientists can tell, by making mosquitoes more sensitive to human odors. That conclusion, though still tentative, stems from studies that show that the dengue virus alters genes known to affect the functioning of the smell sensors in a mosquito's antennae.

These findings provide a very different perspective on vector-borne diseases. Not long ago, the insects involved in all these illnesses were assumed to be independent agents. They were calling the shots, not the germs that hitched a ride on them. But according to this new perspective, the passenger may in fact be the pilot.

Thomas, Lafferty, and other experts on parasitic manipulations think people should be paying closer attention to this phenomenon. It's hard to argue with that view, given the devastating impact of these diseases. Malaria, despite mountains of money invested in vaccines and public-health measures to counter it, still afflicts 214 million people in ninety-seven countries. Dengue fever is soaring faster than any other infectious disease, with about 390 million new cases reported annually, and though once largely confined to tropical and subtropical regions, it now extends to southern parts of Europe and the United States. The parasites behind leishmaniasis and the bubonic plague together account for a few million more cases of disease each year.

Clearly, there's a pressing need for fresh approaches to thwarting these epidemics. A promising avenue of attack might be to sabotage the manipulations that fan their spread. Perhaps, for example, a better understanding of the odors that stir insect pests into a feeding frenzy might suggest how to combat them with a subversive form of aromatherapy — a trap consisting of fragrances more attractive to mosquitoes than the human body's bouquet. Or knowledge of the genes a parasite activates to enhance a mosquito's sensitivity to human odor might suggest a way to block its functioning, cutting the insect off from the world of scents — the bug equivalent of a person being rendered blind or deaf. This much is plain: The better we grasp the nuts and bolts of parasites' manipulations, the more likely we'll be to succeed in throwing a wrench in the works or, better yet, in finding a way to turn their power against them.

Not just medicine but also agriculture could benefit from such expertise. In recent years, millions of citrus trees have been destroyed by citrus greening, a devastating bacterial infection that turns ripening oranges and grapefruits green and bitter and then kills the plants. The disease is rapidly spreading with the help of the Asian citrus psyllid, a bug whose behavior the microbe tweaks to enhance its own dispersal. The bug ingests the bacteria when it sucks up juices from citrus leaves, at which point the insect becomes a far more menacing pest. Compared to uninfected bugs, those that harbor citrus greening bacteria reproduce faster, hop more frequently from tree to tree, and travel longer distances.

Arriving in southern Florida groves in 2005, the bacteria—ferried by the bug—quickly advanced through the state, jeopardizing its $10.7-billion-a-year citrus industry (in some parts of the peninsula, 100 percent of citrus plants are now infected). More recently, the destructive duo has moved across the southern United States to citrus-producing regions in the Rio Grande Valley in Texas and in Southern California.

In response to this mounting threat, scientists are now intensively studying how the greening bacterium communicates its orders to the insect. The hope, of course, is to interrupt the chatter and rein in the infection, a development that could benefit citrus growers around the world whose crops are similarly imperiled by the microbe.

So far, we've focused on parasites that view their host as a taxi to take them to their next destination. But many have very different motives for changing the behavior of the animals that harbor them. These parasites, which count among nature's worst thugs and torturers, deserve our attention. Not only is their means of livelihood fascinating— albeit in a diabolical fashion—but insights into their strategies could make us more alert to ways in which parasites might threaten our own autonomy.

3

Zombified

I N E. B. WHITE'S CLASSIC *Charlotte's Web,* a spider weaves the word *terrific* and other messages into her web in a bid to save a porcine friend from the slaughterhouse. My childhood wonderment at her cleverness came flooding back to me on seeing the photo of a real-life spider whose web — though scarcely the supernatural feat of Charlotte's — was nonetheless strikingly original. The eight-legged creature — a tropical orb spider known as *Allocyclosa bifurca* — had abandoned its normal tight-knit circular motif in favor of a sprawling, freeform style unlike anything I'd ever seen emanating from the arachnid school of design. Asymmetric, with silk threads meeting in a kaleidoscope of angles, it looked like the creation of a spider tripping on LSD.

As it turned out, my impression wasn't far off the mark. The spider had indeed been drugged — not by a scientist, as I'd initially suspected, but by a parasitic wasp (*Polysphincta gutfreundi*). Its tyranny over the spider begins when the wasp seizes hold of the spider and deposits an egg onto its abdomen. As the egg matures into a wormlike larva, it makes little holes in the spider's abdomen through which it sucks out juices. With this dependable source of nutrients, the larva grows rapidly while the spider continues building normal webs and capturing prey. After about a week, the wasp larva starts injecting chemicals that induce the spider to, effectively, build it a nursery. The resulting netlike structure, which bears little resemblance to the spider's usual web, has reinforced lines that will better withstand the strong winds and rains of tropical storms, and thanks to its aerial perch, the developing larva will be kept safe from predators on the ground. Lest a bird or lizard attempt to raid the nursery, the spider even weaves a special decoration that will conceal the parasite's presence.

The spider's reward for all its hard work? Just as it's putting the final touches on the wasp's nursery, the larva kills it, sucks it dry of body fluids, and drops its desiccated carcass to the ground. The wasp larva, which has a single row of stubby legs that are covered at their tips with tiny hooks, then suspends itself from its custom-designed web and creates a cocoon. There, encased like a mummy, it will molt one last time and emerge as an adult wasp.

Numerous parasitic wasps exploit different species of orb spiders in a similar fashion. William Eberhard, an entomologist and arachnologist at the Smithsonian Tropical Research Institute and the Universidad de Costa Rica, discovered the phenomenon in 2000, but he chides himself for not having uncovered the manipulation decades earlier. The strategy is very common, he admits. Now seventy, the silver-haired scientist has been fascinated by spiders since his days as an undergraduate at Harvard, when he took a tedious job in the basement of the university's Museum of Comparative Zoology replenishing alcohol that evaporated from jars holding invertebrate specimens. He hated the work in the dank, dark setting, but the curator of the collection eventually took pity on him, invited him upstairs, and taught him how to catalog spiders according to their kinship. Viewed up close — as he searched for characteristics they had in common — they became objects of beauty, as familiar to him as gems to a jeweler. With an eye finely tuned to discern their morphology and habits, he'd spotted several bizarre-looking webs over the years, ones made by orb spiders in a style as jarring a departure from their normal design as realism is from abstract expressionism. What's more, on closer inspection, he invariably found wasp cocoons hanging from these strange webs. "But I didn't stop to think about exactly how that could have happened," he said. "I sort of had the idea that the spider basically got weakened by the wasp's larva on its abdomen so it just didn't have the strength to build a normal web." Since lots of organisms are sickened by parasites, he dismissed the phenomenon as unimportant.

Eberhard can't recall what prompted him to finally question that assumption, but one day it hit him: *Hey, you idiot! This is interesting!* Indeed, when he turned his full attention to the phenomenon, he was astonished. "I realized that the story of the spider being weak and pathetic and hardly able to move was completely wrong," he said. "It was full of energy and working right along, but making something very different."

As best as he can tell, the larva employs a cocktail of chemicals, some of which act on the spider's central nervous system to alter its behavior, and others (or perhaps a single compound) that poison it when its work is done. In one intriguing experiment, he plucked off the larva just before it was about to kill the spider, and the spider not only made a full recovery but its weaving style gradually returned to normal — only in reverse, meaning the last alterations in its design were the first to vanish. That leads him to believe that as the concentration of the cocktail increases, the behavioral effects become more pronounced — hence, when levels fall upon removal of the parasite, the spider modifies its spinning style in backward order. But that's just a guess, he cautions. And he hasn't a clue how the wasp's chemical-signaling system can be "so incredibly selective in affecting some portions of the host's behavior and not others. These are very, very fine instructions that it's giving to the spider. This is not some general order, like *Climb up* or *Jump in the water.*"

To add to the challenge of determining the underlying mechanism, many different types of orb spiders are parasitized by an equally diverse group of wasps, so there seem to be endless permutations in how such manipulations are executed. And the wasp-induced webs themselves are so varied that Eberhard has no idea what to expect from unusual pairings, as happened when, on a walk through a Costa Rican coffee field, he spotted the larva of a rare wasp clinging to the abdomen of a common spider (*Leucauge mariana*). He placed the spider, with the larva still attached, in a jar, hoping that it would continue weaving in captivity — something many spiders are reluctant to do. To his delight, the prisoner adapted well to its cramped cell and immediately got busy. He'd put a curled piece of paper in the jar — spiders have trouble anchoring their threads to glass — and the creature began attaching its line of sticky thread to a surprisingly large number of points on the sheet's curved inner surface. His surprise turned to shock when he realized what the spider was doing: instead of limiting itself to a two-dimensional, planar construction, as is the custom of this species, it branched out in three dimensions. He'd never seen this kind of spider make anything even remotely like it. The web's lines all converged on a central area, where the spider wove a fine-meshed, striplike platform. And instead of suspending itself from the web — as is the habit of parasitic wasps — the cocooned larva lay sideways on the platform, as if taking a nap. ·

Another type of wasp — a much more prevalent variety — also parasitizes this species of spider. But the spell it casts couldn't be more different. In response to its larva, the spider makes a simpler, more streamlined version of its normal planar web with far fewer spokes radiating from the center and no threads spanning them. Instead of an ornate 3-D structure, the result is a skeletal web completely devoid of the host's trademark circular motif. Each species of wasp, it seems, has its own unique potion for bewitching the spider. They're also brilliant at taking advantage of what a spider normally does and adapting that behavior to suit their own needs. For example, Eberhard said, if a species of spider lives in a sheltered retreat, the wasp might induce it to put a door on the retreat to protect its cocoon. Or if a spider normally weaves decorations designed to camouflage itself, the wasp harnesses that talent to conceal its cocoon from its enemies. Simply put, these parasitic wasps know how to get the most out of a host.

Spiders are by no means the only creatures that need to fear the parasitic wasps' coercive tactics. And drugs are not the wasps' only weapons for gaining the compliance of their victims. *Ampulex compressa,* better known as the jewel wasp because of its iridescent blue-green sheen, performs neurosurgery to achieve its aims. Its quarry is the annoyingly familiar American cockroach (*Periplaneta americana*). Not to be confused with the comparatively diminutive German roach common up north, this species prefers warmer climes and can grow as big as a mouse.

Though dwarfed in stature by its prey, a female jewel wasp that has caught the scent of an American roach will aggressively pursue and attack it — even if that means following the fleeing insect into a house. The roach puts up a mighty struggle, flailing its legs and tucking in its head to fend off the attack, but usually to no avail. With lightning speed, the wasp stings the roach's midsection, injecting an agent that will temporarily paralyze it so that the behemoth will stay still for the delicate procedure to follow. Like an evil doctor wielding a syringe, she again inserts her stinger, this time into the roach's brain, and gingerly moves it around for half a minute or so until she finds exactly the right spot, whereupon she injects a venom. Shortly thereafter, the paralytic agent delivered by the first sting wears off. In spite of having full use of its limbs and the same ability to sense its surroundings as any normal roach, it's strangely submissive. The venom, according to Frederic Libersat, a neuroethologist at Ben-Gurion University in Israel, has turned the roach into a "zombie" that will henceforth take its orders from the wasp and willingly tolerate her abuse. Indeed, the roach doesn't protest in the least when she twists off part of one of its antennae with her powerful mandible and proceeds to suck the liquid oozing from it like soda from a straw. The wasp then does the same thing to its other antenna and, assured that the roach will go nowhere, leaves it alone for about twenty minutes as she searches for a burrow where she'll lay an egg to be nourished by the roach. Meanwhile, her brainwashed slave busies itself grooming — picking fungal spores, tiny worms, and other parasites off itself — providing a sterile surface for the wasp to glue its egg. When the wasp returns, she seizes the roach by the stump of one of its antennae and "walks it like a dog on a leash to her burrow," said Libersat. Thanks to its cooperation, she doesn't have to waste energy dragging the massive roach. Equally important, he said, she doesn't "need to paralyze all the respiratory system, so the thing will stay alive and fresh. Her larvae need to feed five or six days on this fresh meat, which you don't want to rot."

The wasp enters the burrow first, tugs the cockroach in behind her, lays an egg on the exoskeleton of its leg, and then leaves to search for twigs and debris to stop up the opening, thus entombing the fully alert roach. Her offspring then proceeds to clean out the roach's body from top to bottom, at which point the fledgling wasp emerges from the burrow to repeat the cycle.

With the goal of understanding how the wasp dominates her much larger host, Libersat's team fed the winged tyrant a radioactive compound that was incorporated into her venom. When the wasp stung a roach, the researchers could then track where the poison went. They discovered that the venom knocked out a vital neural center for decision-making. Basically, information from the roach's eyes and other sensory organs is relayed to this nexus, and after processing these inputs, the insect chooses what to do next. As Libersat explains, roaches aren't automatons that react to stimuli in the same way every time. Just like us, they can be unpredictable. They think before acting, which is why they're so good at escaping humans chasing after them with rolled-up newspapers. So when the venom knocks out that central command module, the creature is effectively robbed of free will. Instead of running for its life, it's frozen by indecision. All it takes is a little tug by the wasp to get it over its inertia, and off it trots to its death.

Just how the wasp guides her stinger to this critical brain region — a knot of neurons half as big as a pinhead — proved a much tougher riddle for Libersat's team to solve. Her precision, he said, is on a par with medicine's most advanced systems for locating and destroying very small targets in the brain. To coax the wasp into yielding up her secret, the researchers played a trick on her: They removed the roach's brain and then gave the insect to her to see what she would do. The wasp probed the creature's head for nearly eight minutes before giving up in frustration.

That and other experiments eventually enabled them to crack the mystery. The end of the wasp's stinger, they discovered, has special mechanoreceptors that sense tension and pressure. When the stinger reaches the sheath that encapsulates the insect's brain, it hits a slightly resistant material that bends. "That tells the wasp, 'Push here and then spray,'" he said. "It's kind of like a tactile sensation."

As if her state-of-the-art surgical skills weren't impressive enough, the jewel wasp is also a creative chemist. Analyzing the insect's venom, Libersat's team was intrigued to discover that one of its ingredients is dopamine, a neurotransmitter known to initiate grooming in rats. Could the chemical, they wondered, explain how the wasp induces the roach to clean itself of parasites that might harm her larva? Sure enough, injecting dopamine into an unstung roach triggered grooming. Dopamine also has a profound effect on an animal's motivation, offering another hint to how the wasp tames its quarry. "These wasps are way better at manipulating the neurochemistry of their prey than are the neuroscientists who study them," marveled Libersat.

In other corners of nature, parasites manipulate their hosts for very different reasons. A barnacle in the genus *Sacculina,* for instance, seeks to divert a crab's attention from its young to the care and nurturing of the barnacle's own offspring. It may be hard to imagine a barnacle capable of hatching such a plan, much less having the talent to carry it out, but *Sacculina* is decidedly unconventional by the standards of its clan. It does not have a shell and does not attach itself to rocks, seaweed, or anything else. *Sacculina* resembles a bundle of roots that have invaded the soft, fleshy interior of the crab like a metastatic cancer. If any real-life entity fits the nightmarish image of a body snatcher, it's this barnacle.

In its infancy, the parasite is a free-living larva that swims around until, guided by scent, it alights upon a crab. A female *Sacculina* — the larvae come in two sexes — then drives a sharp, dagger-shaped part of her exoskeleton through the crab's thick armor. Through the tip of this weapon, she then injects a tiny, worm-shaped clump of her cells, leaving behind her bulky outer coat. Once inside, her cells grow into a thick tangle of roots, eventually invading the crustacean's eyestalks, nervous system, and other organs. The crab continues to behave like its uninfected kin, trolling the shore and munching on mollusks. But the food it gathers fuels *Sacculina*'s efforts to overthrow its rule, which eventually entails sterilizing the animal, a favorite trick of parasitic manipulators.

The crustacean ceases to mate or grow any larger and henceforth lives only to feed the barnacle and help it reproduce. In the spot where a female crab normally sprouts a pouch on her belly to hold her brood, the colonizer pushes out her tendrils and grows a brood pouch of her own. Guided by smell, two male larvae will find their way to the chamber and begin fertilizing the female's eggs. "In effect these two males become an integrated part of the female," said Jens T. Høeg, an expert on parasitic barnacles. "She becomes functionally a hermaphrodite." As the eggs develop, the crab keeps the parasite's pouch clean by brushing off algae and other parasites, and when they hatch, the crustacean migrates to deeper water. There, it liberates the hatchlings in great pulses and stirs the currents with its claws to send them on their way. And so the barnacle's babies will be carried off by the tide — to commandeer other crabs.

Don't think the host's service to the parasite is over, however. Quite the opposite; it's just begun. *Sacculina* will keep churning out new batches of eggs, and every few weeks, the crab will return to deeper water to repeat the same ritual of dispersing the parasite's offspring. For the rest of the crustacean's life, it will cease to have a will of its own.

It's not just female crabs that are forced into a lifetime of servitude. The barnacle can turn a he into a she. Normally, male crabs have narrow abdomens, but once invaded by *Sacculina,* his body assumes the broader shape of the female's, and it also develops a pouch to hold the parasite's young. Completing this sex change, the feminized male displays the maternal instincts that will make him a tender and protective caretaker of the parasite's brood.

From Scandinavia and Hawaii to the southern coast of Australia, the world is home to over one hundred species of *Sacculina,* and the parasitic barnacle infects a stunning percentage of crabs in certain localities: in Danish fjords, as much as 20 percent; in Hawaii, 50 percent; and in some parts of the Mediterranean, 100 percent. You can identify the infected crabs by the yellow, mushroom-like growths on their undersides — the parasite's pouch. Because infected crabs can no longer molt to grow a larger shell, they also tend to be covered by seaweed and barnacles (ordinary, noninvasive ones, that is). Should you encounter one of these hybrid creatures scurrying along the shoreline, stop to admire the parasite's achievement. That eight-legged form may act like any other crab, but it's really an amphibious robot.

FUNGI WOULD SEEM TO HAVE little in common with parasitic barnacles, but one type seizes control of its host — a carpenter ant — by colonizing its body in a similar fashion. That's not to say its tactics are exactly the same. It doesn't exploit parental affection for its own gain. But its ultimate goal is identical: it wants its host to seek out an ideal spot to spread its offspring and give them a bright start in life.

Even when this fungus, *Ophiocordyceps,* is a mere spore, there's nothing docile in its manner. When it comes into contact with a carpenter ant, the spore sprouts tendrils that burrow into the insect and quickly invade its entire body. It then commands the ant to climb a sapling at exactly solar noon. About one foot up, the insect moves to the underside of a leaf on the northwestern side of the plant and clamps down on its main vein — a sturdy point of attachment. At the same time, the fungus destroys the muscles that control its mandibles, guaranteeing that its jaws will stay forever clenched. Frozen like a statue, the host dies and out of its head springs the fungus — a single stalk that grows a fruiting body at its tip. It soon bursts, spraying spores down onto the ground, where new ants pick them up.

David Hughes, an Irish-born entomologist at Pennsylvania State

University, was the first to document the phenomenon. "Some of my early papers in 2004 and 2006 were rejected because people just didn't believe it," he reported. Zombie ants — as he refers to them — not only exist but are very common.

In rainforests around the world, as many as twenty-two of their grisly cadavers can be found per square yard (or, for those who prefer the metric system, twenty-six per square meter). "I call these dense graveyards the killing fields," he said.

To understand the fungus's quirky commands to the ant, Hughes has moved leaves with zombie ants attached to them to slightly higher or lower elevations or different sides of the plant. The transplanted fungi aren't as successful in propagating themselves in these cases, so clearly there is an evolutionary logic behind the parasite's very precise commands to the ant — but Hughes is hard-pressed to determine what it is. He believes that it might have something to do with the fungus thriving in cool temperatures and very humid air. The northwestern side of the plant gets less sun, and lower leaves are more likely to be cast in shade. How the fungus gets the ant to clamp down on the main vein of a leaf completely mystifies him, however. That's not part of the ant's normal behavior, so "there's no a priori reason to think it would be able to tell the vein from the lamina."

Why it bites at solar noon is also perplexing to him, but he theorizes it might be related to the fact that four hours later, coinciding with sunset, the fungus will be moving from the inside to the outside of the ant, a rough transitional period when it's at greater risk of dying. Nightfall provides the dark, humid conditions conducive to its growth, so it may be trying to synchronize a vulnerable point in its development to coincide with that advantageous time of day.

In addition to his fieldwork, Hughes studies the manipulation in his lab, where he keeps the ants' brains alive in jars (the organ doesn't need a body to function!). To these containers he then adds fungi. As the parasite begins to sprout, it produces a slew of chemicals, a few of which mirror compounds found in the ant's body. But he suspects it may also use foreign chemicals to control the ant — notably, a powerful hallucinogen. His hunch is based on the parasite's close kinship to the ergot fungus from which LSD is derived.

SO FAR, EVERY MANIPULATOR I've highlighted has been a parasite, but not all of them fit that mold. Some manipulators alter the behavior of another creature for their own gain but offer a benefit in return. In that sense, they might more aptly be called symbionts. Since these kinder rulers also challenge the notion that one's thoughts are strictly one's own, I will be including symbionts in future discussions. What's more, we ourselves may be influenced by them: Bacteria that normally live in the human body are suspected of manipulating our behavior to both their and our advantage. We will get to them later, but let's now turn our attention to a benevolent manipulator that involves humans only tangentially but will nonetheless make coffee and tea lovers smile. The story features a drug peddler with petals — that fragrant delight we call a flower.

It begins over a decade ago, when German pharmaceutical researchers discovered that several varieties of flowers laced their nectar with caffeine. Geraldine Wright, an American neuroethologist at Newcastle University in England, came across a report of their finding and was dumbfounded. Caffeine in seeds and leaves is nothing new — it's toxic and bitter-tasting to insects, so plants often use the compound to repel them. But finding an insecticide in nectar was the last thing she'd expected. After all, it's the sweet food that flowering plants use to entice bees to pollinate them. As she read on, however, she noticed that the amount of caffeine in nectar was much less than in other parts of the plant, which suggested bees might not even be able to detect it. For years Wright had been studying bees with the goal of understanding the mechanisms underpinning human learning and memory because at a molecular level, bees' brains are quite similar to our own. So she began to contemplate whether such a low dose of caffeine might affect them like a stimulant as it does for us. Then inspiration struck: Maybe flowers, she thought, were using the drug to improve the insect's memory so that it would return to cross-pollinate them. The bee might also benefit from remembering where an excellent source of nectar is located.

Wright had the ideal skill set to investigate that hunch. She's also, not surprisingly, a pro at handling bees and unafraid of their sting. (Before speaking with her, I went to her university website, where I found a photo of her clad in a "bee bikini": a swarm of the insects discreetly shielding her delicate female parts from view. Bee bikinis and beards are "a macho thing that beekeepers do," she said.) But while she had the know-how and chutzpah to test what caffeine does to bees, she lacked the means. Like many a struggling scientist, Wright had no funding, forcing her to underwrite the project herself. On her own dime, she traveled to Costa Rica to gather nectar from the flowers of coffee plants — the obvious choice to test her theory.

It took two months of labor — time, it turned out, ill spent. Her suitcase containing vials of the substance was lost by baggage handlers at Heathrow airport in London. Again at her own expense, she flew back to the country to repeat the task. ("Don't feel too badly for me," she said. "I had a lot of fun. It doubled as a vacation.")

Several months later, she finally had enough coffee nectar to begin testing her theory in her UK lab. But to her dismay, she found that her bees had superb recall with or without a boost from caffeine. After scratching her head, she decided that maybe the memory test she'd given the bees had been too easy. She'd trained them to recognize a single floral scent and measured how well they could remember it the next day. But in the wild, bees jump from flower to flower every thirty seconds, which means they have to remember hundreds of scents over the span of a day or two. In human terms, she said, "it's comparable to cramming for an exam, which requires you learn a whole bunch of information in very little time, versus studying less information over a longer period of time, when people remember much better." She raised the difficulty of the memory test and struck scientific gold. The bees did abysmally without caffeine, but when they got the normal dose in their nectar, "they performed almost perfectly. It was quite a stunning result. I think this is the first case where we see a pharmacological manipulation of an animal by a plant."

Based on body size, Wright calculates that the insects consume a dose of the drug roughly equivalent to what a human gets from a weak cup of coffee.

Could flowers be manipulating us too? "Probably." Wright laughed. But more, she clarified, as a side effect of evolution. Because the human brain shares common building blocks with that of bees, she believes, caffeine influences our cognitive functioning too. We famously rely on it to stay alert and productive — in short, to give us that buzz we envy in bees. But oddly, it's not clear whether caffeine improves human memory. Some research indicates it doesn't, but when scientists at Johns Hopkins University recently looked at one very specific type of recall not carefully studied before, they found that it may be a memory booster after all. Intriguingly, the drug improves a kind of recollection we rely on for telling apart very similar but different objects, such as types of cars, hammers, or, ironically, flowers.

"It's pretty funny when you think about it," said Wright. "Caffeine is the most widely used drug in the world, and bees have been consuming it tens of millions of years before we showed up on the planet."

There's another amusing twist to her findings. Among the multitudes of flowering plants found on earth, according to her, very few have caffeine in their nectar, yet this tiny minority of plants is among the most widely cultivated today. In addition to the coffee plant, these include tea, cacao (from which we make chocolate), and kola (a popular nut chewed in equatorial Africa). As Wright wryly notes, we like how caffeine makes us feel, so "you could say that these flowering plants manipulate us by getting us to grow vast plantations of them."

During this whirlwind tour of manipulators, I've repeatedly trumpeted their cleverness. But before we move on, it's important to note that manipulators can make mistakes. They can get confused and jump aboard the wrong vehicle, by which I mean a species that cannot further their reproductive goals. The parasite steers itself into an evolutionary dead end. This happens more frequently than you might think. Indeed, manipulators can cause massive collateral damage — an unfortunate fact of life that is commonly overlooked when calculating the full extent of their influence on animal and human behavior.

Which brings us to the cat parasite.

4

Hypnotized

T HE MAN AT THE OTHER END of the line had a thick Czech
accent. His name was Jaroslav Flegr and he was an evolutionary biologist at Charles University in Prague with a very strange
tale. He believed his mind was not fully under his own control. He often felt like an alien force was propelling his actions. That force was
the cat parasite, a single-celled protozoan that he referred to by its
scientific name, *Toxoplasma gondii,* or occasionally as toxo or *T. gondii*
for short. He wasn't sure how he'd been exposed to it. Because cats —
the only species in which it can sexually reproduce — shed the parasite
in their feces, one common way people become infected is by changing the animals' litter boxes. However it happened, he told me, *T. gondii* was in his brain and he strongly suspected it had altered his personality and made him more risk-prone. What's more, his research led
him to think that it might be tinkering with the brains of millions of
other people, contributing to car crashes, mental illnesses like schizophrenia, and even suicides. If you add up all the ways it might harm us,
he said, "it might even kill more people than malaria. Certainly in the
industrial world, that's true."

Flegr sounded paranoid, but there was reason to think he might be of sound mind — or at least, as sane as one could reasonably expect for a man with a parasite in his brain. I'd been pointed in his direction by a rock star among neuroscientists, Robert Sapolsky of Stanford University, whose own team's animal studies certainly showed the parasite was involved in "some pretty wild neurobiology." Flegr's studies, he added, were "well conducted, and I can see no reason to doubt them."

Medical evidence also suggested the parasite might be capable of what the Czech biologist alleged. Since the 1950s, doctors have known that if a pregnant woman becomes infected with the parasite, it can attack her fetus's nervous system and eyes, sometimes triggering miscarriages; if development continues till birth, the child is at greater risk of being born mentally incapacitated or blind. (Please note that if a woman gets infected before becoming pregnant, it poses no threat to her future offspring.) Toxoplasmosis, as the acute infection is called, has also long been known to pose a hazard for people with compromised immunity; they similarly may suffer damage to their eyes or develop encephalitis, a potentially fatal inflammation of the brain. Those at greatest risk are people who are receiving chemotherapy for cancer or taking drugs to suppress the rejection of a transplanted organ. Another well-known group at risk are people with HIV. Especially in the early years of the AIDS epidemic, before good treatments to counter the virus were available, toxoplasmosis was frequently responsible for the dementia associated with the disease. Given *T. gondii*'s demonstrated malevolence and habit of targeting the brain, it did not seem too big a stretch to imagine that it might be stirring up less overt mental trouble as well.

But according to the standard medical wisdom, healthy people who were exposed to *T. gondii* typically developed only brief flulike symptoms, after which the parasite quietly nestled down inside their brain cells and caused no further health problems. It became, in medical parlance, a "dormant infection."

I weighed the competing views and vacillated. Could an eccentric scientist with little name recognition outside of Czechoslovakia really know more than the medical establishment?

In search of an answer, I Googled him. Up popped a photo of a man with hair a startling shade of orange that sprang from his head like wispy peaks of cotton candy. I decided it might be prudent to call a few more experts on *T. gondii* for their take on the scientist. Joanne Webster, a parasitologist at University College London who's widely viewed as a leading authority on the germ, described his work as "controversial," but, she said, "a lot of his studies have been successfully replicated. The standard medical and veterinary textbooks say we needn't worry about the latent phase of the infection, but we shouldn't underestimate this parasite." At the Stanley Medical Research Institute in Bethesda, Maryland, schizophrenia expert E. Fuller Torrey also took the Czech scientist's work seriously. "I think it's entirely credible," he said. In fact, Torrey told me, he himself believes there might be a link between toxoplasma and schizophrenia.

IT HAS BEEN A LONG uphill battle for Flegr to gain any recognition for his ideas. At the time he was embarking on his studies, the Czech Republic was recovering from decades of domination by the Soviet Union, whose apparatchiks tended to promote scientists based on who toed the party line rather than on the merits of their research. As a result, the country was viewed as a scientific backwater, which did not enhance Flegr's reputation. During the Soviet era, there were few opportunities to travel abroad, hence the reason he never mastered English — the lingua franca of science — limiting his ability to disseminate his findings. But by far the biggest obstacle in his path, in his opinion, has been that "people are offended by the idea that some stupid parasite might be able to influence our behavior."

To that list of hurdles, I would add one more: His theory smacks of fringe science. Reviewers likely placed it in the same category as alien abductions, telepathic dolphins, and healing crystals.

What set Flegr down his unusual path of inquiry was a passage he came across in 1981 in the then newly released book *The Extended Phenotype,* written by one of his idols, the British scientist Richard Dawkins. The passage described the suicidal ant that climbs a blade of grass at the command of a trematode in its brain — the very same parasite that a decade earlier had stimulated the young Janice Moore's revelation that manipulative organisms might be a potent force of nature.

The manipulation — the first Flegr had ever heard of — made a huge impression on him. Bizarre as it may sound, it got him thinking about whether a parasite might be at the root of puzzling aspects of his own behavior. For example, he reacts oddly in frightening situations. "I think nothing of walking into the middle of traffic," he told me, "and if a car honks at me, I don't jump out of the way." Even gunfire doesn't particularly alarm him. As a young man, he'd been visiting southeastern Turkey with other students when a battle broke out between the Kurds and the Turkish army. His friends were paralyzed with fear as bullets whistled overhead. He took cover but felt strangely calm. *What is wrong with me?* I thought." Equally odd, he made no effort to disguise his hatred of the Communists, feeling no alarm at what the consequences might be. In hindsight, he said, "I'm surprised I wasn't imprisoned."

Over the next decade, the weird incidents piled up. Though slight of build, Flegr knows karate. Yet if someone attacked him, he would not defend himself. When he realized shopkeepers were cheating him, he didn't speak up. "Something kept me from protecting myself." He started to think that he was somehow being hypnotized by other people — a concern that began to increasingly preoccupy him. With colleagues, he would get into long conversations about whether it was truly possible to be hypnotized, and, if so, how could you prove it? Then, one afternoon not long after such a discussion, a fellow scientist asked him if he'd participate in a research project aimed at improving the sensitivity of a test for toxoplasma. Flegr agreed to be a guinea pig and soon learned that he was infected with the parasite. He immediately began to wonder if it was to blame for his recklessness and fears about being controlled by outside forces.

He dug into the scientific literature. Rats and mice, he learned, typically become infected with *T. gondii* when they come in contact with cat feces on the ground while scavenging for food. Should those rodents then be eaten by a cat, the microbe breeds in its gut and will be defecated out the other end, back onto the ground. And so, in this fashion, *T. gondii* goes round and round. Delving deeper into the literature, he was excited to read that a British scientist named William M. Hutchison had observed in the 1980s that infected rodents were hyperactive. Since cats are drawn to fast-moving objects, Flegr wondered if the parasite was making the rodent run around more than it ordinarily would. In addition, Hutchison found that infected rodents had greater difficulty distinguishing between familiar and novel environments — perhaps, in Flegr's view, another way the parasite might contrive to get the rodents eaten by cats. More ominously, Otto Jírovec — revered as the father of Czech parasitology — reported in the 1950s that individuals with schizophrenia were more likely to harbor *T. gondii*. "The parasite can't know that it's in our brain and not a rat's," Flegr reasoned, "so maybe it's changing our behavior too."

Further research by him revealed that roughly 30 percent of people in the world were walking around with the parasites in their heads — the majority of them completely ignorant of the fact — so if there was any foundation to his suspicions, the health repercussions could be enormous. And he learned that cleaning a cat's litter box is far from the only route by which people become infected. You can get it by eating vegetables that haven't been scrubbed properly or by failing to wash your hands after gardening. Grazing livestock can also pick up *T. gondii* from the ground, and the rapidly replicating organism not only travels to the animal's brain but also produces thick-walled cysts in its muscle — the flesh that we eat. For that reason, people who consume undercooked beef or lamb are at greater risk of acquiring the infection. Indeed, in France, a nation that loves meat *saignant* — literally, "bleeding" — more than 50 percent of the population is infected. (Americans will be happy to hear that rates in the United States are far lower, typically ranging from 12 to 20 percent.) Yet another way humans become exposed is by drinking water contaminated with cat feces — a common occurrence in the developing world, where a staggering 90 percent of people have the latent infection.

To test the manipulation hypothesis, Flegr would have liked to pick up where Hutchison had left off and run more detailed experiments on rodents. But housing and feeding animals is expensive, and, like most Czech scientists in the period immediately following the Soviet era, he was short of funds. So he opted for "cheaper experimental animals — college students." With the goal of teasing out psychological differences between infected and uninfected subjects, he devised questions based on his own intuition about how the parasite might alter behavior or thoughts. Samples included these:

Would you fight till the end if physically attacked?

Do you believe that other people can control you through hypnosis or other means?

Do you respond slowly or passively to imminent danger?

If you realized you were being swindled, would you protest?

To conceal the true purpose of the study from the participants, he mixed the questions randomly with 178 questions on a standard personality inventory.

The results were not what he expected. An individual's infection status had no bearing on how he or she answered any of his ten questions. But people with the latent infection did stand out on a number of traits, and, weirdly, gender influenced their personality profiles. Compared to uninfected men, males who harbored the parasite were more inclined to break rules and were more reserved and suspicious. In contrast to uninfected females, infected women were more likely to abide by rules, and they also tended to have warmer, more outgoing personalities.

Skeptical of his own findings, Flegr subsequently administered the personality inventory to more than five hundred people outside of academia — for example, women who were tested for the parasite during pregnancy, and blood donors. Again he found a similar gender-specific trend in traits associated with the parasite. Participants in one trial were also observed in the lab by a rater blind to their infection status. In keeping with the finding that infected men are inclined to flout convention, the male subjects in this study were the most likely to arrive late for their blood tests and were judged the scruffiest. "They often wore dirty old jeans and rumpled shirts," said Flegr. And infected women? "They were the most punctual of all and dressed the nicest. They wore nail polish, expensive clothes, and lots of jewelry."

On a computerized test, however, infected men and women were surprisingly alike — and very different from people without the parasite. Seated before a monitor, subjects were instructed to press a button the moment they saw a rectangle pop up anywhere on the screen. Those who harbored the parasite had significantly slower reaction times. Closer analysis of the data revealed that infected people's performance began to deteriorate a few minutes into the task as their attention began to wander — an observation that set him thinking about whether they might be a danger behind the wheel of a car. Safe driving, after all, requires constant vigilance and a quick response to changing road conditions. Flegr launched a study of 592 residents of central Prague and found that people with the parasite were 2.7 times more likely to be involved in traffic accidents than age-matched controls free of the infection. Since the strength of epidemiology lies in numbers, he conducted a follow-up study of a much larger population, tracking the incidence of traffic accidents in 3,890 Czech military draftees. Those identified as having the parasite at the start of the investigation went on to have significantly more crashes. "I estimate that as many as a million road deaths a year can be blamed on toxo," Flegr told me.

More than a dozen years into his toxoplasma research, he made another provocative discovery — or, rather, rediscovery. While rummaging through papers buried at the bottom of his desk drawer one day, he came across his first study on the topic, and scanning the tables of data, he noticed he had made a statistical error in calculating his results. Rerunning the numbers, he found that participants' responses to a few of his ten questions were strongly influenced by their infection status after all. In contrast to uninfected people, both men and women with the protozoan were much more likely to believe that they could be controlled through hypnosis. They also were far more inclined to report reacting slowly or passively to imminent threats and experiencing little or no fear in dangerous situations.

IN PERSON, FLEGR IS MORE subdued than flamboyant in style. His office on the top floor of Charles University's natural science building was a sunny, inviting space with a skylight and a window with a tree-top view. Noticeably absent from his desk and floor were the towering stacks of papers that I've come to view as standard décor in the inner sanctums of academia. Flegr was clearly a man of tidy habits, though that was less evident, by his own admission, when it came to his appearance. True to his toxo status, he wore old sneakers, faded bell-bottom jeans, and a buttoned-up shirt that billowed around his waistline.

After shaking hands, I blurted out, "I bet you love cats, right?" I said that mostly facetiously but also just to be sure of how he felt about the animals, as Flegr's thinking did not always follow a predictable line of logic.

His expression immediately softened, communicating an unmistakable infatuation with felines. "At home, we have at least two of them," he said fondly. "They have a little door and sometimes cats from the neighborhood come to visit too." He went to his computer to show me photos of a black-and-white tuxedo cat and a tortoiseshell looking plump and happy in his lap. A boy and a girl — both with red hair — appeared in some of the photos. They were his children.

"You don't worry that they'll become infected?"

"I would not like for my children to become infected. No. But if you keep your house clean, it is relatively rare [that infection will occur]." After the parasite's egglike oocysts are defecated by a cat, "it takes three to five days of exposure to air before they become infectious. Wipe your counters and tables down more frequently than that, and always wash your hands after changing a litter box," he said, "and it's relatively safe."

The word *relatively* kept popping up, but he continued with further reassurances. A cat will produce only a single batch of the parasite's oocysts, he reported. It cannot become infected a second time, so it's only for a brief period once in your pet's lifetime that it could potentially transmit *T. gondii* to you. And not all cats, he added, get infected. In fact, pets always kept indoors cannot get the parasite. "I think that gardening is the leading source of the infection," he opined.

When the conversation turned to his research, his expression became solemn. "It took several years for me to believe my own findings," he said. "Now I no longer doubt, but interpreting the data . . . that is much more difficult." For example, he was puzzled by the finding that infected men and women sometimes have opposite traits. "One possibility is that the parasite may cause stress, and men and women experience stress differently." There is a theory in psychology, he continued. Females cope with anxiety by reaching out to others. "They tend and befriend," is how it's known in the field, he said. Stressed men, in contrast, are more inclined to withdraw.

"Can you guess from observing someone whether they have the parasite — myself, for example?" I asked.

"No," he said, "the parasite's effects on personality are very subtle." If, as a woman, you were reserved before being infected, he said, the parasite won't turn you into a raving extrovert. It might just make you a little less reserved. "I'm very typical of toxoplasma males," he continued. "But I don't know whether my personality traits have anything to do with the infection. It's impossible to say for any one individual. You usually need about fifty people who are infected and fifty who are not in order to see a statistically significant difference. The vast majority of people will have no idea they're infected."

Flegr gave me a brief tour around his lab, which consisted mostly of computers, as his grad students spend a large portion of their time entering data from questionnaires. But of late, he reported, he'd expanded his research to include biochemical measurements and gotten some "very interesting" results. He'd found that male students with the parasite had higher testosterone levels than uninfected males at the university. Female students shown photos of subjects' faces, moreover, rated the infected men as more masculine. He cautioned that these findings were preliminary.

On the way back to his office, we stopped at another lab, where he pointed to a wall with a framed black-and-white photo of Otto Jírovec — the venerated Czech parasitologist who a generation earlier had found a high incidence of toxoplasma infection in schizophrenic patients. Flegr told me he himself had just completed a study on the same topic that exploited brain-imaging technology not available in Jírovec's day. When we were seated again in his office, he handed me a copy of the newly published paper. Only forty-four people with schizophrenia participated in the trial, but small as it was, there was nothing ambiguous about the results. Based on MRI scans, twelve of them had missing gray matter in parts of their cerebral cortex — a puzzling but not uncommon feature of the disease — and they alone had the parasite. I shot him a raised-eyebrow look that said *Yikes!* and he replied, "Jiří had the same response."

Jiří Horáček, a psychiatrist at Prague Psychiatric Center and Charles University, had collaborated with him on the study — but only after Flegr had nagged him for months. As Horáček later told me, he kept putting Flegr off because he was "initially skeptical that *T. gondii* might be behind the reduction in gray matter in schizophrenics. But when we analyzed the data, I was amazed at how pronounced the effect was. To me that suggests the parasite may trigger schizophrenia in genetically sensitive people." (Before anyone freaks out, be aware that only 1 percent of people will be diagnosed with schizophrenia, so given the rampant prevalence of the dormant infection, it's obvious that harboring toxoplasma does not dramatically heighten one's chances of becoming schizophrenic.)

Over recent years, some of Flegr's findings have proven more robust than others. His research showing infected people are more prone to traffic accidents is now supported by two independent studies conducted in Turkey and a third in Mexico. Still another study in Mexico found that industrial accidents are more common in people who harbor the protozoan. And numerous studies by independent researchers continue to link the parasite to mental illness. Weirdly, it turns out that infected male rodents have abnormally elevated testosterone levels, yet Flegr himself was unable to verify the same association in men outside of the student population. "I am no longer so confident that the hormone is elevated in infected men," he said. And when he switched to using a new personality inventory that psychologists have come to view as more accurate than its predecessor, he got completely different results. The sex differences disappeared. Both infected men and women were found to be less conscientious and more extroverted than their uninfected peers.

In spite of these inconsistent or seemingly conflicting results, Flegr still believes the parasite is exerting an influence on human personality. "In laboratory experiments we study genetically identical animals that have been exposed to very similar environmental factors during their lives. Therefore they will probably react identically to toxoplasma infection. In contrast, humans are much more variable in their traits and life experiences and, consequently, in how they respond to the parasite. It must also be remembered that psychology is not mathematics. Any psychologist knows that extroversion and consciousness measured with the Cattel questionnaire [the first test he used] are different personality traits than the traits labeled with the same or similar names that are measured with Big Five [the other test used in his studies]." In other words, according to him, the disparity in his findings may be largely an artifact of the measuring scales he employed.

Flegr could be right, but an alternative explanation is that he's seeing a trend where there is none. The toxo personality is a fiction. In fact, even if Flegr had consistently found the same personality traits linked to the infection with both tests, correlation is not the same as causality. To provide an analogy, it was thought for many years that coffee consumption raised the risk of cancer — until it was discovered that drinkers of the beverage were much more likely to be smokers than non–coffee drinkers were. Once that was taken into account, the association vanished. Flegr may similarly have overlooked an extraneous factor that might undermine his interpretation of the data.

Even though his findings have not been completely reproducible, ongoing research — much of it being done at world-renowned universities — suggests we should not quickly dismiss the notion that the parasite can alter mood and personality or affect one sex differently than the other. Although the landscape is shifting rapidly, strange patterns in the data — eerie echoes of Flegr's own findings — keep cropping up in both animal and human studies. It's too soon to forecast whether history will brand him as a kook or a visionary or both. But there's unquestionably much greater concern that the latent infection may not be so benign after all, with many labs in the United States, Europe, and Asia now intensively investigating the issue.

BY COINCIDENCE, AT THE VERY TIME the Czech biologist first began to suspect that *T. gondii* might be a manipulator, a young British scientist at Oxford University, Joanne Webster, was entertaining the same thought — only she had the resources to test her hunch on animals. Webster would not learn of Flegr's work for many years to come, but when she did, she was struck by the parallels in their discoveries. "I was delighted," she said. "His findings are what we would have predicted based on the animal model."

By the time Webster entered the field, examples of parasitic manipulations in invertebrates were a dime a dozen, but there was a decided paucity of cases in vertebrates such as mammals, making her keen to prove the phenomenon applied to them as well. "That got me thinking toxoplasma would likely be a prime candidate for manipulating behavior. With the parasite sitting in the rodent's brain and the cat its definitive host, this was a natural system where you'd predict it to occur."

She soon confirmed earlier observations of her fellow countryman Hutchison, who'd shown that infected rodents were more active and less vigilant around predators. Then, studying rats in a large outdoor enclosure where they could roam freely, she found something much more remarkable. In one corner of the pen, she placed water; in a second corner, a drop of rat urine; in the third, cat urine; and in the last, urine from a rabbit — an animal that does not prey on rats. She theorized the parasite might reduce the rats' aversion to the cat odor. "We were quite surprised to see that not only did it do that, but it actually increased their attraction. They spent more time in the cat-treated areas," she said. Her team reran the test, replacing the cat urine with urine from other animals that preyed on rodents — for example, dogs in one trial, minks in another. Infected rats were not drawn to the smell of any predators except cats, leading her to dub the phenomenon "fatal feline attraction."

In search of how toxoplasma could pull off this spectacular feat, her team tagged the parasite with fluorescent markers and tracked where it went in the brain of infected rodents. They expected to find its cysts localized in specific parts of the brain owing to its very precise effect on the animal's behavior and the fact that infected rats, save for their attraction to cat odor, appeared to be perfectly healthy. But instead, the researchers found the cysts scattered across the brain — albeit in slightly higher concentrations in some neural centers. That led her to speculate that the parasite fanned out like scatter shot across the brain but induced behavioral changes only when it accidentally wound up in key areas that influenced the animal's emotions or primal drives.

For most of the 2000s, the parasite's mechanisms remained a black box. But in 2009, a parasitologist at Leeds University, Glenn A. McConkey, succeeded in prying its lid open a crack, exposing a hidden talent of the parasite. Ironically, toxoplasma was not initially the focus of his research. McConkey had spent years studying the protozoan that causes malaria. But the one-celled creature happens to be a close relative of toxoplasma, so his team began to compare the DNA sequences of the two organisms on the assumption that any ways in which their codes diverged from each other might provide clues to why they caused very different kinds of illnesses. In the middle of this work, it occurred to McConkey that *T. gondii* alone has a strong affinity for the brain and alters the behavior of rodents, so maybe the microbe had genes that coded for neurochemicals that allowed it to communicate with the animal's nervous system. At the time, many genes involved in the functioning of a mammal's brain had already been identified by other scientists, so he searched those databases for the relevant sequences in toxoplasma's DNA. Up popped such a gene in its genome — but no corresponding gene turned up in the sequence of the malaria parasite. The gene, he was fascinated to see, coded for a protein involved in the production of dopamine, a neurotransmitter that plays a central role in pleasure — think sex, drugs, and rock 'n' roll — and also powerful emotions like fear; in humans, the chemical has been implicated in posttraumatic stress disorder. It also regulates attention and activity levels. "I was shocked that dopamine's functions fit so well with observations of infected rats," said McConkey. They're hyperactive, they're less vigilant, they're less fearful of cat odors. In fact, something about their enemy's scent must be pleasurable, or why else would they be drawn to it? "Then I saw the literature on the dopamine hypothesis for schizophrenia."

For forty years, researchers have noted that people afflicted with the disorder often have elevated levels of the neurotransmitter — another one of those odd observations about the disease, like its tendency to erode gray matter, that have long puzzled doctors. Could the parasite, once ensconced in its host's brain, be churning out the chemical? He and Webster decided to collaborate to find out. By 2011, they had their answer: neurons harboring the parasite were making 3.5 times more dopamine. The chemical could actually be seen pooling inside infected brain cells.

The revelation also brought earlier findings about the parasite into sharp relief. For example, previous researchers had shown that when antipsychotic medicine was added to a petri dish in which *T. gondii* was happily dividing, it stunted the organism's growth. To Webster this suggested the drug might be suppressing the symptoms of schizophrenia by thwarting the parasite. Pursuing the idea further, she infected rats with *T. gondii* and then gave them antipsychotic medicine. Lo and behold, they didn't develop fatal feline attraction. Suddenly, Flegr's claims about the parasite changing human behavior no longer seemed so far-fetched.

Meanwhile, a team of neuroscientists at Stanford led by Robert

Sapolsky was in the midst of their own studies of toxoplasma and watching these developments with keen interest. Parasitologists and neurologists don't tend to travel in the same circles, so Sapolsky didn't learn about Webster's work on fatal feline attraction until the early 2000s. "This is flabbergasting," he wrote in *Scientific American* not long after reading her paper. "This is akin to someone getting infected with a brain parasite that has no effect whatsoever on the person's thoughts, emotions, SAT scores or television preferences but, to complete its life cycle, generates an irresistible urge to go to the zoo, scale a fence and try to French-kiss the pissiest-looking polar bear."

He was equally intrigued by studies linking the parasite to reckless driving. At a salon-style forum organized by the science think tank Edge, he shared that he'd discussed the connection with doctors engaged in toxoplasma testing at an ob-gyn clinic associated with Stanford, and in the middle of their conversation, "one of them jumps up, flooded with forty-year-old memories, and says, 'I just remembered back when I was a resident, I was doing a surgical transplant rotation. And there was an older surgeon who said, If you ever get organs from a motorcycle accident death, check the organs for toxo. I don't know why, but you find a lot of toxo.'"

The British reports about the parasite — coupled with how little was known about its action on the brain and titillating hints that it might ensnare humans in its schemes — made toxoplasma "irresistible to study," said Sapolsky. In 2007, without having secured any funding, he took a bold intellectual leap and joined a small but growing band of researchers vying to explain how a protozoan could seize the reins of a mammal's brain.

His team set out to replicate Webster's work on fatal feline attraction, which they soon did. Like the British group, the Stanford researchers found the parasite was widely distributed in the animal's brain but slightly more prevalent in some areas — notably, centers that are highly responsive to dopamine and associated with fear and pleasure. Several years into the project, Sapolsky reported that their studies suggested the parasite was "basically disconnecting fear circuits.

That helps explain the loss of the aversion to cat odor. But you have to explain something more," he continued. "The rats *like* the smell of cats."

Figuring out why proved much harder to untangle. But gradually, clue by clue, he and his collaborators began to piece together the parasite's modus operandi. *T. gondii* travels not only to the brain but also to the testicles, where it bumps up production of testosterone. What's more, female rats are far more receptive to copulating with infected males. "It's a very strong effect," Ajai Vyas, Sapolsky's postdoc at the time of the discovery, told me. "Seventy-five percent of the females would rather spend time with the infected male." The parasite also invades the male's sperm so when he copulates with a female, it can infect her pups, creating still more vehicles for transporting *T. gondii* back into the belly of a cat.

To say this is bizarre is an understatement. Females of most species are acutely sensitive to any sign that a potential mate may be infected with a parasite — for example, dull plumage or lusterless coat — and scrupulously shun them. But toxoplasma has turned this rule of nature upside down. It also raises a worrisome question: Could toxo be an STD in humans too?

"That's something we're trying to figure out," said Vyas, who has since moved to Nanyang Technological University in Singapore. "We are trying to find some biopsy samples from human testicles."

Just when I thought the science couldn't get any weirder, it did. In the middle of these developments I got a call from Flegr announcing that he had a paper about to be published "proving fatal feline attraction in humans." By that he meant that infected men liked the smell of cat pee — or at least, they rated it more favorably than uninfected men. Infected women showed the reverse trend — they found it more offensive than uninfected women did. The sniff test was done blindfolded and also included urine from a dog, a horse, a hyena, and a tiger. None of the subjects, whether infected or not, showed a significant difference in their reaction to any of the other samples.

"Is it possible cat urine may be an aphrodisiac for infected men?" I asked. "Yes. It's possible. Why not?" said Flegr. I thought he was smiling at the other end of the line but I wasn't sure.

Soon Vyas and I were back in touch, too, because his research was advancing quickly, and he was starting to think that *T. gondii*'s ability to raise testosterone levels played a far more important role in drawing a rat to a cat than the parasite's knack for producing dopamine. Even with no parasite in the picture, he pointed out, a rat puffed up on testosterone is cocky, aggressive, and bold. (A stack of research, incidentally, shows that's true for humans as well. For example, male hedge-fund traders on the floor of the London Stock Exchange were found to make riskier trades on days when their testosterone, as measured from saliva samples collected from them each morning, was highest.) So at the most basic level, he said, the parasite is merely taking advantage of what the hormone normally does.

But one change leads to another, magnifying the mischief. When the excess testosterone produced by the testes reaches the brain, Vyas discovered, it sets in motion a cascade of chemical changes that ultimately alters how the DNA coiled up inside some neurons is read (this is what biologists call an epigenetic transformation). Genes tell cells what chemicals to produce, how much, and when. So the neurons in that part of the brain — a region involved in smell — don't work quite the same as a result. The upshot of all this is that when an infected animal catches a whiff of cat odor, the smell fires up not only those neurons whose job it is to yell "Retreat!" — just as you'd expect — but also neighboring neurons that are activated by the enticing scent of a mate and tell the animal to approach. Put simply, said Sapolsky, "toxo makes cat odor smell sexy to male rats." The confused male often advances only to discover he's courting a cat.

No less wondrous is the female side to this story. When infected, the ladies also experience dramatic hormonal shifts — but of a different sort. The parasite, Sapolsky's team has shown, raises their blood levels of progesterone, which regulates their sexual cycle. And coinciding with that, they start behaving recklessly — in fact, just like infected males jacked up on testosterone. Much remains to be learned about the yin and the yang of the parasite's methods. But toxoplasma never ceases to dazzle, so scientists think it's possible, perhaps even probable, that it has evolved a completely different bag of tricks for manipulating each sex.

Not all of its effects are so specific, however — at least, that's now the view of the Stanford crew. In 2013, Sapolsky retired from doing research, but neuroanatomist Andrew Evans, a member of the group that continues investigations at the university, updated me on their work. He reports that when the parasite invades the brain it can cause lots of different types of mischief depending on where it lands. In some rodents, for example, it can migrate to the hypothalamus — a region that regulates sex hormones and "is a prime site for toxoplasma cysts." The parasites can also cluster in many other brain centers relevant to predator aversion — regions involved in risk assessment, impulse control, spatial memory, and navigational ability, to mention just a few. "We see correlations between specific behavior and where cysts are located in the brain," said Evans. "Natural selection can result in multiple convergent mechanisms that lead to the same result."

The Stanford team made another important discovery: Only half of rats infected with the parasite have cysts in their brains even though all of them have antibodies against it in their blood. The rest evidently fought off the invader before it could proceed to the head. Encouragingly, Evans suspects humans may similarly succeed in blocking the microbe's advance to the brain in many instances.

However, a rat — and, by extension, a human — who loses that battle typically ends up with two hundred to five hundred cysts in the brain, the team's research shows. And each is more than just a potential dopamine factory. Every cyst also triggers a local immune response that further disrupts the balance of neurotransmitters in the surrounding area. Basically, the body tries to starve the parasite of a chemical it requires to awaken from its dormant state, but that very same chemical is needed by the brain for normal mental functioning, so the containment strategy may come at a cost to the host.

"The parasite is going to be altering dopamine, GABA, glutamate, and other key neurotransmitters at two hundred different places in the brain," said Evans, "so it's not surprising it's going to be subtly influencing human behavior" — or, if the cysts happen to cluster in certain regions, even contributing to psychiatric disease. "I do take seriously these reports of an increase in suicide and schizophrenia related to the parasite." It's certainly plausible, he said, that the organism could "exacerbate an underlying mental condition. For instance, we may all exist on a schizophrenic spectrum. Someone without the parasite might already be showing mild symptoms of schizophrenia, but then they get the parasite and their symptoms worsen." The variation from individual to individual in the distribution of the cysts might also explain reports of infected people experiencing personality changes, increased impulsivity, and less fear or poor judgment in potentially dangerous situations such as "when to speed up and pass a car."

Evans cautions that having cysts in the brain does not automatically spell mental trouble for the host. A few hundred of them sounds like a lot until you consider the brain has many billions of neurons that are very good at rerouting signals, allowing information to flow smoothly around diseased regions.

Sapolsky also urges us to keep the potential dangers of the dormant infection in perspective: "I'm not too worried, in that the effects in humans are not gigantic. If you want to reduce serious car accidents and you had to choose between curing people of toxo infections versus getting people not to drive drunk/texting, go for the latter in terms of impact."

Webster takes a similar, if slightly darker, view. "I don't want to cause any panic," she emphasized. "The vast majority of people will have no idea that they've got it. And those who are affected will mostly demonstrate subtle shifts of behavior. But in a small number of cases — we don't know how many — it may be linked to schizophrenia, obsessive-compulsive disorder, attention deficit hyperactivity disorder, or mood disturbances. The rat may live two or three years while humans may carry it for many decades, which is why we may be seeing these severe side effects in people."

AFTER HEARING SO MANY PERSPECTIVES on the parasite, it occurred to me that I had no idea what it actually looked like. So during a visit to Stanford, I asked if I could peer at it down the barrel of a microscope. Evans was away on vacation, but Patrick House, then a graduate student and now a PhD in neuroscience, escorted me to a lab where he set up a slide for me to view.

House explained en route that he'd come to toxoplasma research via philosophy. "I was at UC Berkeley. I was interested in freewill kind of questions. A lot of the philosophy books would approach modern neuroscience but then shy away." Then, while browsing through a used-book store, he came across a volume published in 2003 that had a passage about fatal feline attraction and its potential human implications. "I picked it up, read it, and immediately realized that's what I wanted to study." What made the subject so compelling for him boiled down to this: "Most of us are comfortable with the idea that a painkiller or drug might modify our behavior, but there's something very different about a small parasite — a few hundred or thousand of these single-celled parasites — that's in your brain for your entire life. And because you can't get rid of them and you don't know that they're there, like at what point does all of their influence just become who you are?"

"Are you," I interjected, "even legally accountable for your own actions?" I threw that out there just to be provocative — to test where he thought the field might be heading in a decade or two. But the future had already arrived. Professors at Stanford's law school had recently invited him to address that very issue at an informal colloquium, he told me, and it was clear he was as shocked as I was by their interest. "To see that within the time frame of my PhD, basically nobody having heard of toxo to being questioned by law professors — that's fantastic. That's rapid change."

The lawyers and their students kept hammering him with one question: How do you know if individuals have the parasite in a part of the brain that would influence behavior?

He explained that even in rodents, scientists can predict only how the group will change in response to the infection; they cannot say for certain how any single animal will react. Legal scholars told him that the courts are very strict about accepting evidence obtained from new scientific methods or discoveries, so, given the current limitations of the field, they were profoundly skeptical that a person's infection status would qualify as admissible evidence in most trials.

"Still, they said they might use it — in capital punishment cases," he clarified, displaying excitement tinged with disbelief. When the death sentence is on the table, he was told, the courts are much more lenient about the kind of evidence that can be used in someone's defense. Since the convicted may have to pay for heinous crimes with their lives — making the sentence effectively irreversible — society is willing to lower the bar for what constitutes credible mitigating factors in those cases.

Although experts in criminal law had invited him to speak at the colloquium, health-care professionals also showed up and posed questions he hadn't anticipated: If the latent infection predisposes you to psychiatric illnesses, should it be treated as a pre-existing condition? Should car insurance companies hike your rates on the grounds that you're more likely to get into a crash? "Intellectually," he said, reflecting back on the day, "I like the idea of there being caveats to free will, but I never thought about there being costs to caveats to free will or the practicalities of readjusting the whole infrastructure of society."

The time had come for me to get a peek at the tiny predator — or so I thought. Actually, House explained to me as I bent over the microscope, what I'd be seeing were neurons in a mouse brain that the parasite had infected or interacted with. I was confused by what he meant by "interacted with" — at which point I learned yet another disturbing new fact about the parasite. It doesn't always invade neurons;

occasionally, as it's migrating through the brain, it injects them with a cocktail of chemicals and then moves on. "We call this 'spit and run.'"

"What's in this spit?" I asked. "And what is it doing to brain cells?"

"We don't know, but it's spitting into a lot more neurons than it's invading," he told me, likely magnifying its capacity to disrupt an animal's behavior. Both the infected and the injected neurons glowed neon green due to a fluorescent tag used to visualize them, and according to House, they ranged in number from several dozen to five thousand or so, depending on the mouse. "No two of the mice's brains look alike. And all the animals in the study got exactly the same dose of the parasite at exactly the same age under exactly the same conditions. So when I think that a billion-plus humans have this parasite, I cannot help but believe that every one of them probably has a slightly different infection."

NOT MANY PSYCHIATRISTS HAVE VENTURED into the exotic realm of toxoplasma research, but E. Fuller Torrey is among a handful who have — and he's amazed by how well the animal findings complement the studies on people. "This has not been a fashionable route of research so it almost seemed too good to be true," he told me when I visited him in Bethesda, Maryland, at the headquarters of the Stanley Medical Research Institute, the organization he led for many years, and one of the nation's largest private funders of research on schizophrenia and bipolar disorder. (He currently serves as its associate director of research.)

For almost four decades, Torrey has championed the notion that infectious organisms may be a common cause of mental illness. A personal family tragedy, he admits, may be behind the passion and zeal with which he's pursued this heretical idea. In 1956, completely out of the blue, his younger sister, then a popular high-school senior with plans to attend college, developed pronounced symptoms of schizophrenia. Torrey, a student at Princeton at the time, rushed home to help his widowed mother deal with the crisis.

It was still the dark ages in psychiatry. Leading experts held the view that the disease was a desperate reaction to cold, unloving parents and they implicitly conveyed this impression to his mother, piling guilt on top of her anguish over her daughter's plight. Within a decade, Torrey himself had become a schizophrenia specialist and, finding the dogma of the day highfalutin rubbish, began a dogged search of his own for the disease's causes. He decided to begin at the beginning, scouring centuries-old medical records and historical sources for clues.

"Textbooks today still make silly statements that schizophrenia has always been around, it's about the same incidence all over the world, and it's existed since time immemorial," he said. "The epidemiological data contradicts that completely" — a conclusion he documents in *The Invisible Plague,* a book he coauthored.

Save for the ancient Egyptians, he found, virtually no one kept cats as pets until the latter part of the 1700s. The first people to embrace the practice "were poets — avant-garde, left-wing types in Paris and London, and it just came to be the thing to do." They called it the "cat craze," and coinciding with it, the incidence of schizophrenia rose sharply.

The disease "is so striking in its full manifestations that I find it extraordinary that it wasn't described clearly in the medical literature before 1806, when it was simultaneously described both in England and France," he said. "I mean, there were some very good observers back there in the fifteenth and sixteenth centuries."

Most compelling, people with schizophrenia are two to three times more likely to have antibodies against the parasite than those who don't have the disorder. That's based on an overview analysis of the world literature on the topic — a total of thirty-eight high-quality studies — that he conducted in collaboration with Robert Yolken, a pediatrician and neurovirologist at Johns Hopkins University.

Human genome findings clearly show that schizophrenia has a heavy hereditary component, a discovery that would appear to clash with their position — but Torrey and Yolken don't see it that way. So far, the genes that have been most consistently linked to schizophrenia are those that control how the immune system defends against infectious agents. In families with a high incidence of the mental illness, they say, it may be that the risk factor being passed down is an ineffective immune response to the parasite.

Rubella, Epstein-Barr virus, influenza, herpes, and other germs, Torrey and Yolken believe, likely contribute to the burden of schizophrenia. In addition, the researchers assume some cases involve precipitating factors that have nothing to do with microbes. Heavy cannabis use and birth complications, for example, have also been linked to the disease. But so far, the parasite is one of the strongest environmental triggers identified. "If I had to guess, I'd say about three-quarters of cases of schizophrenia are associated with infectious agents," said Torrey, "and I believe that toxo is involved in a majority of those."

Adding further fuel to the debate, other researchers — chief among them Teodor Postolache, a Romanian-born psychiatrist at the University of Maryland — have now begun to connect the parasite to suicide. While researching suicide risk factors, Postolache made an intriguing observation: People who are profoundly depressed or suicidal are more likely to have brains that show signs of inflammation — an immune reaction to infection or injury. That clue and others led him to think toxoplasma cysts might be an instigating factor in some suicides, a theory he sought to explore with the help of collaborators in Europe. Across twenty-five nations on that continent, he and his colleagues found, the suicide rate among women rose in direct proportion to the prevalence of the parasite in each country. In concert with other researchers, his team also conducted a prospective trial of 45,271 Danish women who'd been tested for toxoplasma at the time of giving birth. Over the next fifteen years, those who had elevated antibody levels against the parasite were 1.5 times more likely to attempt suicide. For those with the highest antibody levels, the risk doubled. Postolache's team and independent groups have gone on to link the parasite to suicidal behavior in both sexes and in places as diverse as Turkey, Sweden, and the Baltimore/Washington area.

"I don't think that at this point we know that toxoplasma leads to suicide," Postolache cautioned. "It could be that mental illness may make you more likely to be exposed to toxoplasma."

Since voicing that opinion, however, he's now more confident that the parasite may be an instigating factor. The reason is a study he and other collaborators conducted on one thousand people randomly chosen from a registry in Munich who were carefully screened to rule out any history of mental illness. The subjects were then asked to fill out a questionnaire to assess their suicide risk and given a blood test for toxoplasma. Compared to the uninfected participants, the infected group was significantly more likely to display traits linked to suicide. These included impulsiveness and thrill-seeking behavior in men, and aggression — both toward others and oneself — in women. Reminiscent of Flegr's findings, a number of the traits correlate strongly with dangerous driving and other reckless behavior.

"I want to see independent groups replicate our findings," said Postolache. "That's going to be critical for moving forward."

Given all the trouble brewing around toxoplasma, should cat lovers be thinking about severing their ties to the animals?

Most scientists agree with Flegr that people don't need to take drastic steps to stay safe from the parasite. In fact, numerous studies show that cats confer many psychological benefits on their owners, so forsaking their companionship might, if anything, worsen rather than improve mental health. Taking care while changing litter boxes, scrubbing vegetables well, and wearing gloves while gardening, say experts, are effective ways to lower the risk of infection. Since beef and lamb are common sources of exposure to the parasite, they also advise cooking meat well or, if you prefer it rare, freezing it first to kill the microbe's cysts. Most important, cover children's sandboxes when not in use. These are favorite sites for cats to bury their feces.

Unfortunately, when prevention fails, there's little doctors can do at present to rout the parasite from the brain, as its cysts have thick walls that make them impregnable to most drugs. Propelled by concerns about the latent infection, however, several groups are now searching for medicines that will overcome that obstacle. Owing to the parasite's close kinship with the protozoan that causes malaria, a major thrust of their research is screening malaria drugs for their effectiveness against toxoplasma's cysts. In tests on mice, the strategy has already identified a few promising agents, raising hopes that human treatments for the latent infection may follow.

To explain where we are in this endeavor, Yolken draws an analogy to medicine's stance toward ulcers in previous decades. "People for years had suspected that *Helicobacter* was causing ulcers but we only knew for sure when we had good treatments against the bacteria. So that's what we need. The ultimate goal is to show that when we get this out of people, they get better."

Postolache cherishes the same hope, of course. And while it may be premature, he admits that findings from the vanguard of the field have already got him thinking differently about human behavior. "Many times we don't know why we do what we do," he said. "We usually associate mood disturbances with conflicts in early childhood, but who knows? Some of our unconscious may be controlled by pathogens."

Unfortunately, toxoplasma most likely does not have a monopoly on the manipulation of our minds. As we'll see, there are other parasites that may influence many of the elements that are central to our sense of self — our moods, appetites, recall, and reasoning abilities.

5

Dangerous Liaisons

T HE IDEA FOR THE EXPERIMENT sprang out of a casual conversation. Janice Moore had been invited to give a talk on parasitic manipulators at the State University of New York at Binghamton. The day before the event, Chris Reiber, a biological anthropologist there, had picked her up at the airport as a favor for another colleague and invited her back to her home for dinner. Reiber didn't know Moore or her research very well, so as she prepared food, she peppered her with questions. Hearing Moore's stories about manipulators in nature immediately made Reiber think of sexually transmitted diseases in people. Before coming to Binghamton, she'd had a job at a neuropsychiatric institute at UCLA that worked closely with nearby clinics treating patients with HIV. "The clinic directors used to tell me that HIV-positive patients go through these terrible end-stage phases where they had intense cravings for sex," Reiber told me. These were anecdotal reports and not well documented, but she'd started to wonder if they might have a basis in fact because when she attended scientific conferences, other health professionals shared similar stories. Maybe, she suggested to Moore, these urges, if true, were the virus's attempt to spread before the host's imminent death.

It was an interesting notion, Moore agreed, but there was no way to prove it without doing something no one would dream of: infecting healthy people. You had to be able to compare behavior before and after the exposure to make a strong case, she said.

"So how would you explore this in humans?" Reiber pressed her.

Soon the two scientists were batting around ideas about other types of microbes that might want to manipulate us, germs that would be safer to study. What about a cold virus? Maybe it makes you more social in order to fan its spread. Why not expose people to a mild cold virus? But again, they nixed the plan as too risky.

Then Moore got an inspiration: "Doctors give people the flu all the time," she said. By that, she meant they gave them flu vaccines, which contained all the same molecules found in the live virus except for its dangerous infectious component. Moore hypothesized that the inactivated flu virus in the vaccine would induce the same behavioral changes in its human hosts as its untamed twin did. Tracking people's social habits before and after receiving the vaccine, they both agreed, might be a relatively easy and ethical way to show that parasites manipulate humans, overcoming a major criticism of Flegr's studies of people: the oft-repeated charge that correlation does not establish causation.

After Moore returned home, the two scientists began exploring in earnest how they might conduct the trial. Delving into the medical literature, they discovered that the flu virus is most transmissible in the two to three days after a person's exposure to it but prior to the onset of symptoms. Indeed, viral shedding peaked during that narrow window of time. Put another way, if you go to a party and wake up the next morning with a sore throat and runny nose, don't assume the people you hugged or shook hands with the night before gave you the bug. Quite likely just the opposite happened: you gave it to them.

Once you begin coughing and blowing your nose, you'll probably take to your bed and lie low, providing the pathogen with fewer chances to meet new people. By then your immune system will have kicked into high gear, squelching the virus's ambition. Putting this all together, Moore and Reiber predicted that the germ would prod people to seek out the company of others early in the infection, before it had blown its cover and triggered a counterattack by defense cells.

Once they'd formulated a hypothesis, they decided it would be wise to conduct a pilot trial to see if the idea had any merit. The researchers tracked the social interactions of thirty-six people — none of whom knew the purpose of the study — before and after they got flu shots at a health clinic on the Binghamton campus. The change in the subjects' behavior was huge, so notable that its magnitude surprised even Reiber and Moore. In the first three days after vaccination, coinciding with the time when the virus was most contagious, subjects interacted with twice as many people as they had before being inoculated. "People who had very limited or simple social lives were suddenly deciding that they needed to go out to bars or parties or invite a bunch of people over," reported Reiber. "This happened with lots of our subjects. It wasn't just one or two oddities."

Unfortunately, like many scientists with exciting early findings, they never did succeed in getting the funding to do a larger trial, which would have included a control group that got a sham vaccine. Until that happens, they can't rule out an alternative explanation for their dramatic results; people who get vaccinated may mingle more because, as Moore puts it, they perceive themselves as "bulletproof" — that is, immune to infections.

The idea that pathogens responsible for STDs stoke sexual appetites also remains unproven, but Reiber and Moore aren't alone in harboring such suspicions. On a blog sponsored by the University of Chicago Press for the purpose of promoting the free exchange of ideas between luminaries in science, the Columbia virologist Ian Lipkin wrote, "It is possible, though I have no experimental proof, that when herpes simplex virus infects the sacral ganglia [nerves at the base of the spine], it may (in)advertently stimulate nerve endings in the pelvic area, promoting sexual activity and increasing the likelihood it will move into another host."

In a more recent conversation with me, Reiber speculated that when the herpes virus awakens from its dormant phase to cause genital blisters, it may do more than just rev up someone's libido; as part of its reproductive strategy, it may also fuel his or her desire to have sex with different partners. Like Lipkin, she has no data to support that view, but given the remarkably broad range of talents displayed by manipulative parasites, it's a hypothesis, in her opinion, that bears consideration.

At the University of Montpellier, Frédéric Thomas raised yet another possibility: "A parasite doesn't need to increase your motivation [for sex] because most animals have plenty of motivation. At first opportunity, they act on it. The problem is to increase your attractiveness" when you're most infectious. According to him, as women approach the fertile period of the menstrual cycle, there's evidence that their voices become more animated, lilting and slightly breathless, which makes them seem more excited and engaged in a conversation — a come-on to the opposite sex. "I wouldn't be surprised if parasites are doing something similar."

Weirdly, an infectious agent not traditionally thought of as an STD — the rabies virus — can trigger sudden shifts in libido. A sharp rise in sexual craving, arousal, and pleasure are atypical but well-documented manifestations of the disease in humans. In earlier centuries, the French referred to these uncontrollable desires in women as *la rage amoureuse* and *la fureur utérine*. Men may experience prolonged erections and ejaculations as often as hourly, sometimes accompanied by orgasms — symptoms so dramatic that they were noted even in the ancient world. The Greek physician Galen in the second century recounted that a porter in the throes of rabies had multiple involuntary emissions over three days. Of course, rabid animals can't tell us how they're feeling, but prolonged erections are similarly a sign of the disease in dogs, who may begin furiously humping anything in their path.

Since no discussion of neuroparasitology would be complete without addressing rabies, let's take a closer look at what the virus does. The fact that an infection spread by bare-fanged aggression can occasionally excite the loins certainly gives rabies a lurid appeal. But there is a more pressing reason the disease deserves our attention. Although those blessed to live in societies with excellent health care often think of rabies as a scourge of the past, the disease remains at epidemic levels in poorer parts of Africa, Asia, and elsewhere.

A familiarity with the ravages of rabies will make you profoundly grateful for the efforts of every researcher who's worked to prevent the disease, all the way back to Louis Pasteur, who in 1885 harvested saliva from the fangs of a rabid dog in order to develop the first vaccine. Before then, the only treatment for the much-feared infection consisted of cauterizing the site of an animal's bite or amputating a mauled foot, hand, or limb. Such drastic measures often worked owing to the pathogen's slow mechanism of action. After entering through pierced skin, it doesn't infiltrate the bloodstream, the modus operandi of almost every other virus. Instead, it creeps along the nerve fibers, moving at a plodding pace of a few inches a day, until it reaches the brain, usually about two to four weeks later — though, baffling scientists, the incubation period can in some instances stretch for many months or even a year or more.

For most people, the first symptom is a flulike malaise — an indication that the infection has reached the brain. In short course, the virus typically invades the limbic system, a neural center that controls such fundamental drives as aggression, sex, hunger, and thirst. This is when victims may experience momentous sexual stirrings. As the virus madly replicates, causing circuits to fire erratically, light, noise, a scent, or the slightest touch — even a breeze — can trigger profound agitation. This phenomenon, called hyperesthesia, may have a purpose in common hosts like dogs, raccoons, bats, and foxes: an excitable animal is easily provoked to snap its jaws. The virus also paralyzes muscles in the throat. When people cry out in pain, they emit hoarse choking sounds sometimes likened to a bark. As swallowing becomes more difficult, saliva rich in the infectious agent builds up in the mouth and becomes frothy, spilling over the lips in long threads of drool. Hydrophobia — literally, "fear of water" — often seizes hold of human victims at this point. *Fear*, however, doesn't adequately describe the torment water evokes in the afflicted. Because the virus causes extremely painful contractions of the throat, the sight of any liquid in a glass or splashing in a basin can actually elicit a gag response.

As the disease progresses to the so-called furious phase, victims' expressions may assume a menacing grimace as facial muscles go into involuntary spasms. Unlike rabid animals, people rarely bite, but they may fly into a rage. Terrifying hallucinations are not uncommon at this stage. Death usually occurs a few days after the onset of severe symptoms — typically due to asphyxiation or cardiac arrest.

The disease does not always take such a dramatic course, however. In a third of cases, inexplicably, the sole manifestation of the infection is paralysis, which starts at the site of the bite and gradually spreads throughout the body, leading to coma and death. This is a less violent but usually slower path to the grave, so it has the drawback of prolonging suffering.

As unnerving as the symptoms are, perhaps more unsettling is the fact that this pathogen doesn't need to turn its hosts into savage beasts in order to spread. Before an infected animal begins behaving bizarrely, the virus has already reached high concentrations in its saliva and can be spread when the creature licks specific parts of another animal's body — notably, the smooth pink mucous membranes that line the eyes, lips, mouth, nostrils, nipples, anus, and genitals. "The virus is effecting its transmission by normal mammalian behavior," said rabies expert Charles Rupprecht. "We're social. We love to lick, we love to suck, we love to bite. Sucking is part of the maternal bond. Most mammals are doing a lot of genital licking and sniffing. Dogs do that all the time. Puppies jump up and do play-biting because they're trying to be fed by the mom. During copulation, a male will bite a female on the nape of the neck to subdue her. We're all keyed up on the peculiar things the virus does to us, but by and large they're inconsequential to its spread."

Fortunately for us, we are not as susceptible to the virus as other mammals are. The chances of human-to-human transmission of the virus — for example, through oral sex, kissing, or a love bite — are remote. The few cases that have been reported are anecdotal and come from poor countries where health officials don't have the resources to prevent the disease, much less conduct carefully controlled epidemiological studies to determine the veracity of such claims. Still, Rupprecht is adamant that anyone who's had sex with a person with rabies, or even shared a cigarette or drink with him or her, should be given prophylactic treatment — typically a series of four injections in the arm, which, contrary to myth, are no more painful than ordinary flu shots. "Based on everything we understand about the pathophysiology of this disease, it's certainly feasible that individuals can be transmitting rabies by kissing and oral sex," he underscored. Since a sharp rise in libido can be one of the infection's earliest symptoms, moreover, there is the danger that a rabid person may unwittingly spread the virus before he or she is diagnosed.

In a case reported in India, for example, a twenty-eight-year-old married woman suddenly wanted sex so often that it became a source of marital stress, leading her to consult a gynecologist and another physician and, ultimately, go to an emergency room. There she developed hydrophobia, causing the doctors caring for her to immediately suspect rabies. When questioned, she recalled that two months earlier, a puppy had given her a tiny bite that seemed insignificant at the time. She died in the ER the next day, but her husband was vaccinated and did not get rabies. "In a given milliliter of saliva there's probably upwards of a million virions [virus particles]," said Rupprecht. "For a disease with the highest fatality of any infectious agent, who's going to take that risk? I can't treat you for rabies once you're ill."

In sum, the rabies virus appears to improve its transmission by essentially bludgeoning the brain, causing numerous circuits to go haywire all at once. Some of the symptoms it induces — for example, hydrophobia — appear to be irrelevant to its spread, but heightening sexual arousal may be useful (at least in animals). Certainly inciting an animal to bite by stirring up rage and hyperesthesia — jangling its nerves in response to the least sensation — are methods it uses most effectively for host-hopping. Nonetheless, the fact that rabies can be transmitted before symptoms emerge by a friendly lick or nip on the neck during normal sexual coupling suggests the furious stage in reality serves as an insurance policy — in essence, the parasite's backup plan for finding a new host should it fail to get around by more mundane means.

That's the modern take on rabies. In earlier times, before anyone knew about viruses or how they spread, people no doubt saw the disease as a contagious madness. A savage beast bites you, and through the wound, its spirit enters your body. Thus possessed, you too become a wild animal. You foam at the mouth, rage at the world, and may even, in your delirium, bite. You bark like a dog and display uncontrolled carnality. Violence, sex, blood, and gore. An evil that spreads and has a life of its own.

If that profile sounds familiar, it's because rabies almost certainly provides the foundation for vampire myths. In many of these legends, especially versions originating in Eastern Europe in the first half of the eighteenth century, vampires are people (including, sometimes, the deceased) who rise at night and, often taking the form of a dog or wolf, violate their neighbors — gorging on their flesh, sucking their blood, or raping them, among other heinous acts. Actually, these were not just stories to people living at that time; they were believed to be true and anyone accused of having such savage powers could be hanged or burned at the stake. Count Dracula, penned into existence by Bram Stoker in 1897, built on these ancient accounts, except that his villain famously morphs into a bat. It is surely no accident that these supernatural forms possessed the viciousness and hypersexuality of rabid animals and wore the mantle of rabies's most common hosts — or that vampirism, like the viral disease, can be transmitted by a bite. Nor does the resemblance stop there. In an article published in 1998 in the normally staid pages of *Neurology*, the Spanish physician Juan Gómez-Alonso pointed out other, less obvious parallels between vampires and rabid animals. According to folklore, the lifespan of a vampire was forty days, which coincides with the average time a victim lives after being bitten by a rabid animal. And just like rabid people, vampires were repelled by light (hence their nocturnal habits), strong smells (the odor of garlic, according to folktales, could ward them off), and water (pouring it around graves was recommended to keep them in their underground vaults).

WE'VE EXPLORED THE THESIS that parasitic manipulators may take advantage of our sociality and sex drive. Let's now look at a parasite that might threaten our minds in a very different way. It's not as garish or nightmarish as rabies by any stretch. What makes it concerning to scientists, in fact, is its stealth. They fear it could be quietly eroding the intellect of infected people. And it, too, can be transmitted to us by our cherished pets.

That parasite is toxocara. For those of us who love dogs or cats or both, it may be thought of as toxoplasma's evil twin. A six-inch-long worm, toxocara comes in two species, *Toxocara canis* and *Toxocara cati*, which, as their names suggest, infect dogs and cats, respectively. A strong clue that toxocara could spell mental trouble for us is the fact that the parasite's larvae can become lodged in the human brain. At least the *canis* variety has been shown to do that; less is known about the *T. cati* species, which may not have as strong an affinity for the brain. An estimated 10 to 30 percent of people in North America and Europe are infected with the worm's larvae, and as much as 40 percent of the populations of some poor countries. With numbers like that, you'd think the medical literature would be overflowing with papers exploring its effects on human mental and physical health. But toxocara appears on the U.S. Centers for Disease Control and Prevention's list of the top five neglected parasitic diseases, and experts commonly use words like *cryptic* and *enigmatic* to describe it. Why the snub? Decades ago, the parasite fell off doctors' radar screens because, unlike *T. gondii,* the infection is much less likely to cause serious disease.

Parasitologists, in contrast, have long been wary of toxocara. At

Trinity College Dublin, Ireland, Celia Holland, a leading investigator of the pathogen, worries that it's causing subtle cognitive deficits that have long been overlooked.

TOXOCARA IS BETTER KNOWN to pet owners as roundworm — those pale yellow wriggling threads that dogs and cats sometimes cough up or that are passed in their stools along with thousands of the parasite's microscopic eggs. When the eggs are consumed by another dog or cat, they develop into fast-moving larvae that invade many different organs of the body. Those that reach the gut hatch into worms and begin pumping out eggs, repeating the cycle. The infection is also passed down from one generation of animals to the next, for the larvae that stay behind in other tissues are activated when a female becomes pregnant, at which time they may cross her placenta or pass into her milk, invading her litter.

Roundworms spread to other hosts in much the same way as *T. gondii*. The eggs may be picked up by rodents, rabbits, moles, birds, and other small creatures that in turn may appeal to canine or feline taste buds — providing yet another route by which the parasite can get back to its spawning ground inside our pets' guts. Livestock can also consume the eggs, so people may become exposed to the larvae by eating undercooked meat. The most common way we come in contact with the parasite, however, is by poor hygiene. Most vulnerable in that regard are young children playing in the dirt or in sandboxes contaminated with cat or dog feces.

When toxocara eggs hatch into larvae inside a human body, they do not grow into adult worms; that can happen only in a canine or feline host. Instead, the parasite's development remains arrested at its highly mobile larval stage, allowing it to wander far beyond the gut to organs such as the liver, lungs, eyes, and, occasionally — no one knows how often, given the dearth of research — the brain. In keeping with the parasite's reputation for being innocuous, blindness, seizures, and other severe neurological symptoms are uncommon complications of the infection. But starting as far back as the 1980s, clues began to surface in the medical literature that it might be destructive in more devious ways.

In an early study conducted by Holland and physician Mervyn Taylor in Ireland, two hundred children who had tested positive for the parasite were divided into three groups based on levels of antibodies (a measure of the severity of the infection). Behavioral disturbances, headaches, and disrupted sleep, along with twenty other physical symptoms ranging from asthma to stomach pain, all increased in direct proportion to the youngsters' antibody levels. Other small studies—including two carried out in the United States—compared the cognitive skills of infected children with those free of it. Worse outcomes—for instance, lower academic performance, hyperactivity, and heightened distractibility—were reported in the infected children. And an epidemiological survey conducted in France—the only one to Holland's knowledge that has focused on an older age group—connected the infection to a higher risk of dementia.

All these studies had only a few hundred subjects each, making it impossible to draw firm conclusions from them. What's more, toxocara disproportionately affects the poor, so socioeconomic factors muddied interpretation of the evidence. In short, this smattering of data, while concerning, was hardly persuasive.

A study published in 2012, however, rigorously controlled for those confounding variables—and it bore out Holland's suspicions. The report, which appeared in the *International Journal for Parasitology,* was based on an analysis of a huge body of epidemiological data collected in the United States by the CDC.

Its authors used a battery of psychometric tests to assess the cognitive functioning of a nationally representative sample of almost four thousand youngsters from six to sixteen years old, roughly half of whom tested positive for the parasite. Compared to age-matched controls, the infected kids got significantly lower scores on every measure, including mathematical ability, reading comprehension, verbal digit recall, visuospatial reasoning, and IQ. Impressing Holland, the study's results held up even after the researchers controlled for socioeconomic status, education, ethnicity, gender, and, most important, blood levels of lead, whose toxic effects on the nervous system are well known to depress children's performance at school.

The survey also revealed that toxocara's toll is far from evenly distributed among ethnic groups. Twenty-three percent of African American children were infected, in contrast to 13 percent of Mexican Americans and 11 percent of whites. This implies that disadvantaged minorities may do worse at school not only because of well-known factors like poor nutrition and inferior education but also, possibly, because of parasites in their heads — or so believes Michael Walsh, an epidemiologist at the State University of New York, Downstate, in Brooklyn and the lead author of the study.

Like Holland, Walsh was drawn to toxocara research because of the infection's neglected status. "The parasites that make headlines," he said, "typically are big killers or are gross to look at and make sensational pictures in films." No one pays much attention to the ones that act slowly and stealthily and whose symptoms are less dramatic. While these infectious agents may not maim or murder, they can often harm far more people by keeping their nasty doings hidden. And if the population they target is underprivileged, with poor access to health care, the culprits can easily go undetected.

Toxocara's connection to poverty is multifold. The risk of soiled objects being placed in the mouth increases when children are unsupervised — a situation more common in low-income families, for whom childcare is a luxury. A trip to a vet to get a pet dewormed can also be too costly for those on tight budgets. Making matters worse, playgrounds and green spaces are rare in run-down inner-city areas and hence prone to becoming heavily contaminated with dog feces infested with the eggs.

Poor rural settings aren't safe from the infection either. Stray dogs and cats are frequently infested with roundworms and they live close to people who often feed them, so backyards in such areas can also become major repositories of toxocara eggs.

Neither Walsh nor Holland present their findings as definitive. As

Walsh explained, "We are just waking up to the fact that maybe this isn't such a benign parasite after all. We're in a very similar situation to investigators who were looking at cognitive harm posed by lead in decades past." Once a common ingredient of paint, lead has insidious long-term health effects, though doctors for years overlooked the danger that confronted children who ingested paint chips or inhaled the powdery residue that old paint sheds as it weathers and cracks. Yet the hazard was far from trivial. Even very low doses were shown to retard cognitive development and depress IQ.

It's too soon to say if toxocara is exacting a similar toll in lost intellectual capacity. At this stage, said Walsh, "we're just looking at what on average is happening to the four thousand kids in the study." Some children's immune systems will likely be able to eradicate the infection or prevent the larvae from proliferating in the brain, he believes. In other kids, the immune system may not function as well, or they could be reexposed to the eggs multiple times while playing outside. Toxocara in those youngsters, he said, might depress cognitive development far more substantially.

Paralleling the findings of studies of children, mice fed the worm's eggs typically display difficulties learning new tasks compared to those free of the parasite. For example, Holland found that they drank less from a water bottle hidden in a maze that they'd previously navigated, suggesting that their memory was impaired. Yet they were clearly thirsty, for they were keen to drink when they were returned to their cages. And even though infected animals were as active as controls, they appeared less curious; they didn't show as much interest in novel stimuli or exploring their surroundings — essential traits for staying alive in the wild.

Inspired by Joanne Webster's famous fatal-feline-attraction experiments, Holland ran trials to assess how animals infected with toxocara behaved in situations in which they were at high risk for predation — for instance, when they were placed in an area treated with cat odor or in a brightly lit space. Her results were murky — not a slam dunk as in the case of Webster's toxoplasma findings. An examination of the brain tissue of infected rodents yielded more insights. In the process of attacking the roundworm larvae, their immune systems appeared to be damaging surrounding tissue — and that self-inflicted injury, she suspects, might be responsible for the memory impairment of the infected animals. But there were also hints that the worm's larvae might be capable of the kind of subtle mischief that is a hallmark of a manipulator: the larvae tended to accumulate in the white matter of the mouse's brain, especially in two regions involved in learning and memory. That suggests the parasite may benefit from meddling in those regions and fits well with the observation that infected mice had impaired recall and were less interested in exploring, reducing their familiarity with their surroundings. "The parasite wants to get into the final host and you could argue that some of these effects would definitely make them more vulnerable to predation by a dog or cat," said Holland.

Perhaps the best way to settle the argument is to declare the very premise of the debate absurd. As pioneer Janice Moore would remind us, a behavior induced in a host can be the result of both a pathology and a manipulation, so it may be impossible to distinguish one from the other.

However the issue is resolved, more research on humans is clearly needed to pin down how often the parasite invades neural tissue and the potential risks — especially to children, whose rapidly developing brains are more vulnerable to environmental insults. Unfortunately there's currently no safe and efficient way to detect the worm's larvae in the brain. A CT scan can indeed visualize them, but it would expose children to harmfully high levels of radiation. An MRI gets around that problem, but that diagnostic method is far less accurate at identifying the larvae. In fact, it's even difficult to isolate them by autopsy — the approach Holland uses in her animal research. She typically spends an hour or more per mouse brain, and that's a mere walnut-size nugget as opposed to the three-pound human brain.

In spite of that limitation, both Holland and Walsh believe solid research can bring the potential risks of the infection into better focus.

One strategy Walsh has proposed is to get a baseline reading of the cognitive functioning of a large group of children who don't have the parasite and then follow their progress over several years. When some become infected, he'd look to see if their mental ability had deteriorated since their pre-infection assessment and also in comparison to the youngsters who remained free of the parasite.

In the meantime, he's already thinking about ways to counter the parasite's spread in the New York area, which, if successful, might serve as a model for other communities. His first order of business will be to determine where sources of the infection are located. To that end, he hopes to join forces with the New York City Department of Health and Mental Hygiene to embark on an onerous and odorous task: a massive block-by-block survey of dog excrement throughout New York's five boroughs. Once problem neighborhoods are identified, the goal would be to target them for interventions, such as encouraging pet owners to be more fastidious about picking up after their dogs, educating them on the importance of deworming their animals, and, perhaps, if evidence continues to incriminate the parasite, even subsidizing the cost of treatment. Anthelmintic medicines — deworming agents — for treating toxocara are very cheap and can often prevent a dog from becoming infected if given annually in the form of a pill tucked into its chow. To get owners to cooperate with that treatment regimen, one scheme health officials are contemplating is to make the medicine available at pharmacies — at least for prophylactic use — saving people the added cost and inconvenience of a visit to a vet. And of course, if the carrot doesn't work, there's always the stick — notably, stiffer fines for offenders. To catch violators of pooper-scooper laws, sanitation workers might be enlisted as eyes on the ground. Walsh is even thinking about employing social media such as Twitter to get local residents involved in patrolling their neighborhoods.

His zeal to purge every pooch of the parasite stems from a stark reality: treatment of people probably won't reverse the damage done by the parasite, only, with luck, stop further harm.

• • •

WHAT ELSE MIGHT BE TOYING with our minds?

More than fourteen hundred parasites prey on humans — and that's only the ones that we know about. Untold others await discovery. We have no idea how many, already named or still to be christened, may be manipulators.

Yet, unsettling as that thought may be, we should not read too much negativity into the unknown. Rapidly accumulating research suggests hordes of microscopic manipulators are everyday residents of our bodies, and they don't all wish us ill by any means. In fact, some — the sort that might more aptly be called symbionts — may actually lift our spirits and offer other mental benefits. How we feel and act matters to them. After all, they have a stake in our survival.

6

Gut Feelings

WOULD YOU GO BUNGEE JUMPING? Strike up a conversation with a stranger? Eat a whole pie in one sitting? Strange as it sounds, your gut bacteria may be influencing these and many other choices and habits.

We celebrate the brain as the seat of human intellect, but mounting evidence suggests our behavior is governed not just from the top down but also literally from the bottom up. And for all we know, microbes that inhabit our genital tracts and noses may also have a say in our behaviors and actions.

The study of the human microbiome — all the tiny organisms that call the body home — is a frontier as wild and untrammeled as any in science, and how these tenants influence the brain is perhaps the least understood of all their roles. The majority of them are so supremely adapted to life inside us that scientists have had little luck coaxing them to grow in petri dishes. Only recently have researchers possessed the technology even to estimate their numbers, much less characterize precisely what they do.

The first major census of our microbial tenants got under way in 2005, aided by the development of superfast gene-sequencing machines for distinguishing the genetic fingerprint of different organisms. Spearheaded by an international consortium of scientists, the project initially focused on characterizing microbes that inhabit healthy people. Individuals with acne, bad tooth decay, or more serious ailments were excluded from the survey. From this winnowed-down pool of recruits, the researchers sampled material from stools, armpits, behind the folds of the ear, the back of the throat, between the toes, inside the vagina, and every other nook and cranny that could be reached with a probe. The microbes were then cultured and their genetic material analyzed segment by segment. From the output, computers calculated the size of the microbiota community — the sum total of viruses, bacteria, fungi, protozoa, and other organisms that dwell within each of us. The final tally came to more than one hundred trillion organisms, dwarfing the population of human cells by a factor of 10. The amount of genetic material of microbial origin surpassed our own by 150 times. To put it plainly, 90 percent of you is not you.

Some of these microbes cross the placenta and take up residence inside us while we're still in the womb. But the biggest wave of colonization happens at birth. After the mother's water breaks, microbes lining her vaginal canal jump aboard the infant with each contraction. From that moment forward, each of us becomes a magnet for microbes. They come from the doctors who deliver us, our swaddling blankets, our first pacifiers, and the air around us. They invade every crack and crevice of our bodies, especially the gut, which draws them with its bounty of nutrients. Over the first two years, this population shifts dramatically and is highly idiosyncratic from baby to baby. But then, as infants transition to solid foods, it stabilizes. Children and adults typically harbor on the order of a few thousand species of microbes, and no two people have exactly the same composition of them. Your microbiome is as unique as your fingerprint.

Our microbial cells — or selves, for each of us is in reality a superorganism — defy easy classification. A number of strains labeled pathogens, the census revealed, actually inhabit us all the time, causing trouble only when we're run-down or when unusual conditions favor their growth. The same species of bacteria may be a helper (a symbiont), a harmless freeloader (a commensal), or a hurter (a parasite), depending on circumstances that are constantly in flux.

In the gut, resident microbes take a share of every meal you eat, but in return they aid in digestion, synthesizing vitamins and disarming dangerous bacteria that you ingest. They also churn out virtually every major neurotransmitter that tunes our emotions — notably, GABA, dopamine, serotonin, acetylcholine, and noradrenaline — as well as hormones with psychoactive properties. To varying degrees, scientists now suspect, intestinal microbes influence whether you're happy or sad, anxious or calm, energetic or sluggish, and, by signaling the brain when you've had enough to eat, perhaps even whether you're fat or thin.

Scientists are still trying to figure out how exactly gut bacteria get messages delivered to the distant outpost of the head, but they have a few ideas.

Some psychoactive compounds made by gut bacteria, they believe, are detected by the enteric nervous system — a thick skein of neurons that runs the entire length of the gut. This network has more neurons than the spinal cord — hence its nickname, "the second brain" — and it connects to the big brain upstairs via the vagus nerve, a major route by which gut bacteria make their voices heard. Indeed, 90 percent of information transmitted by this cable goes from the viscera to the brain, not the other way around, as scientists for years had assumed.

In the process of breaking down food, gut bacteria also produce metabolites with neuroactive properties that may stimulate the same nerve cable or be transported by the bloodstream to the brain.

Intestinal bacteria may engage the immune system, which can lower our mood and energy level, yet another pathway by which our microbiota might change our behavior. Perhaps related to that observation, depressed people tend to have abnormally high amounts of certain gut bacteria, and they are more likely to have elevated biomarkers for inflammation — an immune-mediated response.

Intriguingly, certain GI disturbances — notably ulcerative colitis and Crohn's disease — are marked by disruptions of the gut microbiome, and these illnesses are associated with an unusually high incidence of mental disturbances in comparison to serious diseases that afflict other parts of the body. Indeed, 50 to 80 percent of those who suffer from these conditions are clinically depressed.

More surprising, specific abnormalities in the composition of the human microbiota have been linked to autism spectrum disorder (ASD) — a condition characterized by increased anxiety, depression, and impaired social ability. Rodents that display behaviors that mirror many of those observed in children with ASD show similar changes in their gut microbiomes, and when healthy bacteria are introduced into their intestines, their behavior is largely normalized. That observation has raised hopes that it might be possible to develop microbiome-based therapies for ASD — though scientists are a far cry from translating the finding into clinical treatments.

Obviously a lot is going on in the gut. Its microbiota, combined with the immune cells and second brain, constitute a complex and diverse ecosystem — a rainforest growing within each of us, if you will. So figuring out what gut bacteria are doing to one another or to us and why is beyond anyone's grasp. Standard models of manipulation may not apply here. It's clear, however, that animals' behavior changes markedly with the composition of their gut microbiota.

The most striking proof of this comes from germfree mice — that is, animals specially reared under sterile conditions to be devoid of gut microbes. A normal healthy mouse with its intestinal microbiota intact is a quick and eager learner. Show it a new object like a napkin ring and it will circle and sniff it with great interest. If placed in a maze, it's keen to explore new passages. Germfree mice demonstrate none of this natural curiosity. It's as if they have no recollection of objects and places they've recently explored, for they're just as likely to favor what's familiar over what's new, exciting, or different. These rodents are also oddly fearless. They boldly venture where mice with normal microbiota know not to go. Bright lights and open spaces that scream danger to the average mouse don't deter them in the least. In fact, they're so immune to anxiety that they show no signs of distress even when separated from their mothers at birth for three hours a day — a trauma that would normally lead to lifelong skittishness and social maladjustment. In addition, germfree mice scurry about their enclosures more than those with normal microbiota. Just as in the case of mice with autism-like features, transferring healthy gut microbiota into the germfree animals can normalize many of these behaviors — for example, they become more cautious and less active — but only if done before they are four weeks of age. After that, the transplant has no effect, suggesting that microbiota at the start of life shape the very wiring of the brain. Indeed, studies conducted at the Karolinska Institute in Sweden show that early exposure to gut microbiota dramatically affects the expression of hundreds of genes, many involved in transmitting chemical messages in the brain.

Bacteria in the gut may even influence personality. The same team that did the maternal deprivation study — Stephen M. Collins, Premysl Bercik, and colleagues at McMaster University in Ontario, Canada — explored this possibility using inbred mice with two distinctly different temperaments. The mice in one strain were unusually calm and not inclined to socialize with their peers. The mice in the other strain displayed traits at the opposite end of the spectrum; they were more high-strung, aggressive, and gregarious. These two breeds also possessed different microbiota, so the researchers decided to see what would happen if they took germfree mice of one strain and colonized them with the gut bacteria of the other. The result was essentially a personality swap. The calm mice became more agitated and outgoing; the mice that were aggressive quieted down and became less sociable. In other words, each group developed a disposition a little more like the strain that had donated the microbes. Coinciding with that change, the Canadian team detected increased production of a neurochemical called brain-derived growth factor in a part of the brain involved in regulating emotion.

• • •

TRANSPLANTING ENTIRE MICROBIOTA ECOSYSTEMS into rodents is not the only way to study the connection between microbiome and mind. A red-hot area of research uses probiotics — such as the healthy bacteria in yogurt — to the same end. This is the domain of John Cryan, a neuroscientist at University College Cork, Ireland. As a PhD student in the early 1990s he became fascinated by psycho-neuroimmunology — a then-emerging field devoted to understanding how the immune system and brain talk to each other. His interest later shifted to the gut because mounting research suggested it might be at the center of that conversation. Landing a job at a university in Cork furthered his burgeoning interest, as it allowed him to collaborate with what's now known as the APC Microbiome Institute, a research center in the same city whose prime mission is understanding the role of gut bacteria in all aspects of health. After Cryan talked with Ted Dinan, a psychiatrist there, the two scientists decided to join forces to explore the interaction between the brain and the gut.

In one early trial, they subjected healthy young animals to stress — a reliable way to turn them into anxious adults — and discovered that their intestinal bacteria were very different from those of animals raised under kinder circumstances. They also were intrigued by the intimate association between digestive disorders and depression. Then, as their thinking evolved, said Cryan, it occurred to them that "if your gut can cause maladaptive behavior, could it also be responsible for producing some of the better behaviors that might be involved in emotion and learning?" At this point, they got the idea to run behavioral tests on mice fed probiotic bacteria (full disclosure: the lactobacillus culture they used was provided by a supplement manufacturer).

In one such experiment, mice who got the dietary intervention and a control group of untreated animals were placed in a cage and repeatedly given foot shocks paired with a sound. Both groups reacted the same way: they immediately froze — an adaptive response. But the next day, when the researchers played the sound alone to see how well the mice were able to associate it with the punishing stimulus, those given the probiotic froze more frequently than their untreated peers. "They attended better" to the shock, said Cryan. "The animals on lactobacillus learned much better and more efficiently."

Cryan, Dinan, and colleagues also examined how probiotic-fed mice responded to a test widely used by the pharmacological industry to assess the efficacy of medicines used for anxiety and depression. The animals were placed in a small tank of water and forced to swim with no escape route. Panicked, they eventually gave in to despair and became immobile. Untreated animals went only two minutes before they lost their will to survive. In contrast, mice fed probiotics before the test swam a full forty seconds longer. "We were really struck by the magnitude of the effect," said Cryan. "They behaved as if they were already treated with an antidepressant."

Because a neurotransmitter called GABA plays a central role in suppressing the body's response to fear and hopelessness, the researchers looked at the parts of the mice's brains most influenced by that chemical. They found marked changes in those areas. Next, they cut the vagus nerve, the main highway connecting the gut and the brain. The outcome was striking: The animals fed lactobacillus gave up the struggle to survive just as quickly as untreated animals. Not only that, but their brains no longer showed changes in the areas most affected by GABA. "Somehow the bacteria are affecting the vagus nerve," said Cryan. Once it's severed, "signals can no longer get from the gut to the brain to change its neurochemistry."

Findings so dramatic are usually met with skepticism, but quelling doubts in the scientific community, a colleague of Cryan's at McMaster University, John Bienenstock, was able to duplicate the results. The findings also shed light on an often effective but poorly understood treatment for severe, refractory depression. Called vagus nerve stimulation (VNS), it does exactly what its name suggests. Tiny electrodes are attached to a section of the nerve that runs through the neck, and when they're activated by a battery, the brain gets more stimulation than it ordinarily would, usually accompanied by a lift in mood. "This is totally speculative," said Cryan, "but are the bacteria acting as a kind of vagus nerve stimulator"—essentially mimicking the effects of the therapy?

The latest discoveries beg an obvious question: Could probiotics help millions of people who are debilitated by serious mood disorders?

Cryan's team and other groups in Europe and North America are now conducting clinical trials to test the therapy on people whose primary problem is anxiety, depression, or bipolar disorder. The results are not yet in, but studies of groups whose mental woes may originate in GI troubles offer encouragement. In one study of thirty-seven patients with functional gastrointestinal disorders (an umbrella term for irritable bowel syndrome and other common gastric complaints that can't be tied to an underlying abnormality), for example, probiotic treatment not only improved their symptoms but also brought about a significant reduction in their depression and anxiety based on both self-reporting and measurements of stress markers in their saliva and urine. The outcome was impressive given that all of the subjects had previously been treated—without success—at multiple medical centers.

A sprinkling of clinical investigations also suggest that probiotic remedies can soothe colicky babies, a condition that torments 20 percent of newborns along with their frazzled, sleep-deprived parents. In one trial the approach reduced crying and fussing by 70 percent.

In addition, gathering evidence hints that supplements of healthy bacteria may help to buffer already high-functioning people against everyday stresses and strains. For example, a randomized, double-blind trial in France on fifty-five people with no history of psychological disturbances found that regular consumption of a probiotic reduced blood levels of stress hormones and subjects' ratings of their depression, anxiety, and ability to cope—improvements not seen in the control group. Owing to the small size of the study, one should be cautious about extrapolating too broadly from it—scientists laud the investigation mainly for pointing the way toward fruitful avenues of research in the future—but its findings mesh well with a first-of-a-kind brain-imaging study that linked changes in people's neural functioning to a probiotic-rich diet.

Conducted at UCLA by a medical team led by Emeran Mayer and Kirsten Tillisch, the trial involved sixty women — all healthy and with no psychiatric disorders — who were randomly assigned to three groups. One group was given bacteria-containing yogurt twice daily for four weeks, one group was given a nonfermented milk product twice a day for four weeks, and the third group received no intervention. Before and after the experiment, all the women underwent an MRI scan that measured their brain activity during an emotion-recognition task: they had to match photos of faces showing the same expressions, either anger, fear, or sadness. Compared to the other women in the study, those who ate the bacteria-containing yogurt displayed subdued brain activity in three regions involved in emotion, cognition, and the processing of sensory information.

"We have to be cautious in how we interpret the results," Mayer, the study's senior author, told me when I visited him at his sunny office in a massive hospital complex on the UCLA campus. But in his view, the changes seen on the MRI scans suggest that the dietary intervention had a positive impact on brain functioning. "The women who got probiotics were less reactive to negative emotions like anger, fear, and sadness. It might be beneficial to not react as strongly to negative emotions. A lot of sensitive people get very nervous if someone frowns on them." It would be interesting, he added, "to see what would have happened if we'd used happy faces in the trial."

Mayer grew up in Munich and speaks English with an ever-so-faint German accent that lends him an air of European worldliness and sophistication. His most prominent trait is his far-ranging curiosity, which has led him to seek intellectual adventures and to pile up degrees. While still in medical school, he spent several months living with the Yanomami people of the Amazon — photos of whom appear on his office wall — a group whose microbiota he'd love to study. Over the course of his career, Mayer has acquired expertise in an impressive array of specialties that include gastroenterology, psychiatry, and physiology. In a Venn diagram, probiotics would be represented by the space where these disciplines overlap, so, at least to me, his focus on this topic made perfect sense. But that's not how he saw it.

"Seven or so years ago this was a field that I thought was completely bogus," he told me. "I've been contacted many times by probiotic companies and always turned them down. But finally they said, You can design any study you want. And I said, I'll do a very high-risk, placebo-controlled study and I don't think anything will come of it."

But as we saw, the trial didn't turn out as he'd expected — and when he talked about it, he sounded like he was still struggling to wrap his mind around the results, as if he couldn't quite believe his own findings. One reason is that the circuitry involved in the emotion-recognition task "is very hardwired in our brains. Even monkeys have it." Your reaction to people's expressions "happens within milliseconds: Are they angry? Should you get into fighting mode? Are they happy? In which case you should get into the affiliative state. It was very surprising that something as robust as that could be modulated by four weeks of exposure to probiotic species."

You might think the company that supplied the bacteria culture for the trial would be pleased with its outcome, but "they're paranoid that this could affect sales in a negative way — that when people realize, Wow, this affects your emotional state or your mind, they'll stop buying yogurt."

Mayer and many of his colleagues want to decrease their dependence on the probiotics industry for funding to do their work. The industry is reluctant to make its data public, and its involvement creates the perception — correct or not — that the researchers' findings may be biased by financial interests.

These scientists are at last getting their wish, as money from government agencies has begun pouring into the field, giving researchers the means to pursue avenues of inquiry purely for medical reasons, not just because the results might aid some company's marketing campaign.

Presently, the probiotics in foods like yogurt are chosen by companies based on consumers' taste preferences and other commercial considerations, so trying to harness them for therapeutic purposes, said Cryan, is "like going into a pharmacy and picking random pills hoping that they'll affect the brain in a beneficial way." That's not a very efficient method of finding clinically useful bacteria, so many researchers are now adopting more targeted strategies. They're screening probiotic bacteria — including strains not on the market — to determine which ones produce neurotransmitters and other psychoactive compounds. Should promising strains be identified, the next step will be feeding them to rodents, and then, if a reduction in anxious behavior is observed, proceeding to human trials. That process will be far more protracted in the case of noncommercial strains of bacteria because the FDA requires that they be as rigorously vetted as new drugs. Alternatively, it may be possible to dispense with bacteria altogether and create medicines simply from the psychoactive chemicals they make.

Other scientists are sifting through our microbiota with the goal of gaining a better understanding of individual variations in its content — knowledge that could prove important in customizing treatments for patients, whether they suffer from mood disorders, GI disturbances, or both. Though each person's gut microbiome is unique in its composition, scientists have discovered that different species tend to cluster together in individuals, albeit in slightly different proportions. Researchers liken individual microbiomes to gardens. All of us have many bacteria in common — what scientists call the core microbiome. These species are frequently the most abundant. Think of them as being like wildflowers — say, dandelions, poppies, and primroses. The rest of the microbial species typically vary from person to person. To extend the garden analogy, you may have one type of flower — say, marigolds — in large numbers, while someone else may attract purple heather or daffodils. And each of us may harbor rare exotic species — the equivalent of an orchid found only in the highlands of Papua New Guinea. Even the microbes distinct to each of us, however, are alike in one important respect: they tend to fill similar ecological niches, like breaking down different types of protein, fiber, and fat.

Diet — in particular, whether you get most of your calories from fat, as is typical of people who live in places where food is plentiful and cheap, or from grains and vegetables, as is common in agrarian societies — may influence what microbes blossom inside you. You're also likely to have many of the same species your mother does, and not just because her bacteria were probably the first to colonize you. Our genes, it appears, play a role in determining how hospitable our bodies are to different microbes.

Perhaps the most amazing discovery in this area — one that Mayer's group brought to light — is that the architecture of an individual's brain correlates with the types of species that are most dominant in his or her gut. From an MRI scan of your head, said Mayer, "we can actually predict what microbial gardens are growing within you." These species influence the brain's gray-matter density and volume as well as the white-matter tracts that link different regions of the cerebral cortex. In particular, gut bacteria appear to have the biggest influence on the wiring of the brain's reward center, the part that motivates you to seek pleasure and avoid pain. To Mayer, that suggests that gut bacteria might influence "background emotions, stress reactivity, whether you're optimistic or pessimistic. Indians of Venezuela jungle" — like the Yanomami he once lived among — "are exposed to completely different microbes than people living in a city. Obviously they behave very differently too. We don't know if gut bacteria might be related to that."

Needless to say, he'd like to find out.

Gut bacteria could have at least one other very important effect on behavior: they may stimulate food cravings. In the next chapter, I'll explain why they may be motivated to manipulate our appetites and how, with luck, we might enlist their help to win the battle against obesity.

7

My Microbes Made Me Fat

BEHOLD TWO MICE. One is pleasantly plump, the other skin and bones. Yet the thin mouse is by far the bigger eater of the two. It weighs less because, unlike its chubby counterpart, it has no microbes in its gut. Without these helpers to break down its food, most of it passes through its intestine undigested. Though the animal consumes 30 percent more food than the bigger mouse, it has 60 percent less fat.

Studies of germfree mice leave no doubt that microbes have a big impact on the amount of nutrients that we can derive from food — an obvious way they control hunger and body weight. But there's more to this story than simply calories in and calories out. Intestinal bacteria regulate hormones your own body makes to stoke or suppress your appetite — for example, ghrelin, the molecule that goads you to get a second serving at the buffet, and leptin, which tells you to push your plate away. It's also suspected that gut bacteria may themselves synthesize chemicals that signal brain regions governing satiety. These areas include circuits rich in cannabinoid receptors — the same neuropathways that are involved when a cannabis user gets the munchies.

Inspired by these insights, many scientists are gambling that our microbiomes may hold the secret to overcoming obesity. At the center of this flurry of activity is Jeffrey Gordon, a medical researcher at Washington University in St. Louis, who has performed some of the most creative and provocative experiments in the field.

In 2006, Gordon's team made an important discovery: fat mice had a far larger proportion of one major division of gut bacteria and less of another, while thin animals displayed the reverse profile. Obese and thin humans, to Gordon's fascination, showed the same pattern. Did that mean certain bacteria were making people fat? Or could it be that the excess calories consumed by fat people favored the growth of those strains?

To untangle cause and effect, Gordon, along with Vanessa K. Ridaura and others, performed a series of experiments that riveted the scientific community. They launched a search to identify rare sets of twins in which one was overweight and the other skinny (the idea was to minimize the effects of heredity). The researchers then collected bacteria from the subjects' feces and used it to colonize germfree mice that were genetically identical. The animals that got the bacteria from the overweight twins became obese and those that received the bacteria from the thin twins remained slender. Next the researchers staged a battle between the two types of microbes by housing both sets of mice in the same cage. Rodents are coprophagous — a polite way of saying that they eat one another's droppings — so allowing the two groups to mingle exposed all of them to the fecal bacteria originally collected from both the fat and thin twins.

The microbial wrestling match culminated in a dramatic upset: The obese rodents lost their excess weight as bacteria from the thin twins muscled out their founding population. The lean rodents stayed lean. The bacteria from the slim twins prevailed in both groups.

Because all the subjects in this trial were fed standard rodent chow that was low in fat, the researchers wondered what would have happened to the animals if they'd been fed the equivalent of junk food at the time they were exposed to the bacterial mixture. Would the outcome have been the same? The researchers consulted various dietary charts and tables to create food pellets for the rodents that were similar in composition to the sugary, high-fat fare consumed by a large sector of the public in affluent nations. On this diet, obese rodents did not lose weight. Their fattening microbes triumphed over the slimming ones. The lean rodents, however, did not become rotund no matter how much they ate. Their founding population of bacteria protected them from obesity.

More recently, Gordon's team has discovered that fat mice have impoverished microbiomes in comparison to those of thin mice, which harbor far more species of gut bacteria. Related research suggests the diverse microbiota of slim mice can extract more calories from the same food, so you'd think these mice would be the ones ballooning in size. But paradoxically, their bacteria break food down into metabolites that seem to act like appetite suppressants and energy boosters, so the animal burns off the excess calories and consumes fewer overall.

These results hold out the fantasy — and that's all it is for now — that overweight humans could slim down if the right cocktail of bacteria were somehow introduced into their guts in combination with a brief period of adhering to a low-fat regimen to oust the fattening bugs from their systems. Then, when the colony of the good guys was well established, they'd lose their craving for chocolate mousse cake. Or maybe they could eat three servings of it in a row and still fit into skinny jeans.

Alas, it probably won't be that simple. In humans, obesity is a complex disease, affected not just by obvious players like diet, heredity, and exercise but also by sleep habits, stress, cultural norms, romantic trouble, income, smoking, drinking alcohol, pet ownership, and who knows what other factors. Having said that, gut bacteria could turn out to be tipping the scales very strongly in the wrong direction for some people.

With the goal of rectifying this imbalance, scientists in Amsterdam are employing a strategy that entails transferring feces from a thin person into the intestine of a fat person using a colonoscope (the tube-shaped instrument that is inserted into the rectum to perform a colonoscopy). Nasty as this procedure sounds, use of the technique is not unprecedented in medicine. Called fecal transplantation, it's already proven to be promising in experimental trials to treat a variety of GI disturbances, including *Clostridium difficile*, a condition marked by chronic diarrhea and abdominal pain, and Crohn's disease (the donor bacteria come from healthy people with no GI disturbances). That fact emboldened the researchers in Holland to embark on their trial involving obese patients. The move, however, was deemed premature by many experts in the United States, who would like to see the underlying science behind the strategy better pinned down first. The approach also raises safety concerns.

Although donors are rigorously screened to make sure they don't have HIV, hepatitis C, or other infections, there's always the chance that pathogens may go undetected, sickening the transplant recipient. Leaders in the field are quietly discussing whether donors should also be screened to rule out mental illness. Noting that gut bacteria may influence emotions and perhaps even temperament, John Cryan, the University College Cork neuroscientist, warned — only half in jest — that you should be careful who your donor is. He or she "could turn you into something that you're not." Stephen M. Collins — who did the personality studies on rodents — has similar concerns: "Fecal transplantation can be a lifesaving treatment for people with digestive disorders, so one would hate to dampen enthusiasm by suggesting that the procedure could change patients' personality. But there are theoretical grounds to suggest that it might have some impact on behavior." In fact, he reported, his team is currently assessing whether patients who receive fecal transplants for digestive problems experience mood changes as a result of the procedure.

Meanwhile, Gordon and his colleagues are working to identify which bacteria in human feces are responsible for protecting against obesity in their animal trials. If they're successful, the idea would then be to transfer only those purified strains, rather than actual feces, into humans — an innovation that might reduce adverse side effects and cast the procedure's yucky image in a more appealing light. Not just anal but also oral delivery systems might be possible — for example, pills that you swallow ("crapsules," as scientists are fond of calling them) or foods, from baby formula to yogurt, infused with slimming bacteria, perhaps even species tailored to an individual's unique gut microbiome. Another option is to enhance meals with prebiotics — ground-up fiber from root vegetables like chicory that basically serve as fertilizer for your healthy microbiota. As Gordon told *Scientific American*, "We need to think of designing foods from the inside out."

As we move forward on one front, however, we move backward on another. Antibiotics, by depleting gut microbiota, may be expanding the ranks of the obese. That warning comes from Martin J. Blaser, director of the New York University Human Microbiome Program and author of *Missing Microbes,* a book that makes a strong case for this view. No one — certainly not Blaser — is suggesting we do without these lifesaving medicines, but he and other scientists do urge us to rethink how quickly we reach for them. The notion that antibiotics might make you heftier is certainly not novel to farmers. For decades, they've been adding low doses of antibiotics to animal feed to fatten up just about any kind of livestock, from poultry to swine to cattle. (Lagging far behind most European countries, the United States did not begin to phase out this use of antibiotics till 2014.) Another farmer's trick: begin dosing animals when very young to maximize your profits — if you wait till later, the practice is not as effective in building bulk.

We humans start consuming antibiotics early in life as well. Many of us receive our first dose before we're even born. In the industrial world, one-third to one-half of women are treated with antibiotics during pregnancy. By age eighteen, the average American has had ten to twenty courses. We don't down antibiotics with every meal, though, and unlike livestock, we typically take them in higher amounts for shorter spells. So is it fair to extrapolate from animals to humans?

To answer that question, Blaser's group gave young mice brief, high-dose pulses of the drugs to mimic human treatment. When the rodents grew up, they not only weighed more but also had considerably more fatty tissue than their untreated counterparts. And if the animals were fed calorie-rich fare instead of normal rat chow, weight gain was even more dramatic. This synergistic interaction between the foods we eat and our microbiomes, Blaser believes, may go a long way toward explaining why the incidence of obesity in the United States is greatest in the South, which combines a love for deep-fried foods with the highest usage of antibiotics in the nation.

A decade-long study that tracked 163,820 participants from childhood into the teen years also supports the notion that use of these drugs is making us fatter. By age fifteen, it found, kids who had been prescribed seven or more courses of antibiotics were three pounds heavier than those who had never taken any. Though weight gain associated with these medications was modest in size by the end of childhood, the effects appeared to be cumulative over time. This makes Brian S. Schwartz, who led the study at Johns Hopkins Bloomberg School of Public Health, think the difference in weight between the two groups might become far more substantial by middle age. "Your body mass index," he warned, "may be forever altered by the antibiotics you take as a child."

Ironically, a celebrated medical achievement — the banishment of the ulcer-causing bacterium *Helicobacter pylori* — may be a major contributor to some people's expanding girth. The organism plays a key role in regulating ghrelin, a hormone that decreases as your belly fills with food, signaling you to put down your fork. In the microbe's absence, however, levels of the hormone fall more slowly, encouraging you to overeat. Though often cast as a villain, *H. pylori* does not cause problems for most people, according to Blaser. Just a century ago, it was the most common bacteria in the stomach and almost everyone had it, he reports. But today, in wealthy regions of the world, it has largely been wiped out. Only 6 percent of children in the United States, Germany, and Sweden presently have it in their stomachs.

Of course, no one would risk reintroducing this evicted tenant back into the stomach given its potential to cause ulcers and even, if the lesion is allowed to fester for years, trigger cancers of the stomach and esophagus. But there may be a way to capitalize on its virtues while minimizing its dangers, according to Barry Marshall, the Australian scientist who shared the 2005 Nobel Prize in Physiology or Medicine for his role in discovering the bug. He believes that a probiotic of the future might include a detoxified version of *H. pylori* that would offer its hunger-suppressing benefits without its penchant for corroding our stomach lining.

The decrease of *H. pylori* is but one of many trends over the past century or so that have benefited humans generally but may also have hindered their microbiomes. Cleaner water and improved hygiene and sanitation have winnowed down the pool of bacteria in our guts, as has the downsizing of families. Kids today have fewer siblings, who can be counted upon to smother them in kisses, cough in their faces, and steal bites from their Popsicles, depriving many youngsters of microbes that are a healthy fit with their hereditary makeup. Babies delivered by C-section into the sterile hands of a surgeon have fewer encounters with their mothers' microbiomes in the critical early period when gut populations are being established. And newborns fed formula miss out on hundreds of microbial strains in their mothers' breast milk. Possibly related to these developments, studies show that these two groups of infants are more likely than other children to become obese.

Microbiome research may not only provide one more health rationale to breastfeed but also change how babies delivered by C-section are welcomed into the world. In a trial under way in Puerto Rico, doctors are dabbing the skin of such newborns with gauze soaked in their mothers' vaginal fluids. Over the coming years, the investigators will compare the health and weight of these children with those who were born by C-section but who did not receive the bacterial bath.

Better focusing the firepower of antibiotics will also be critical for preserving the variety and vigor of our microbiota, according to Blaser and other scientists. Instead of indiscriminately blasting good and bad germs alike, as we now do, the goal would be to go after the enemy with more finely targeted antibiotics, reducing the collateral damage. In the meantime, we can all protect the gut's ecosystem simply by using these drugs more sparingly and choosing old-fashioned soap and water over germicidal hand lotions and home-cleaning products. Recall that microbes make up 90 percent of our bodies, so zapping every germ in your path is in a sense anti-human.

Until microbiome science yields better treatments for obesity, people desperate to lose weight may be tempted to reach for the probiotic pills and powders now sold at pharmacies and health-food stores. Unfortunately, that may be a better way to lose money than pounds. The promising results I've highlighted from animal and human studies — including research demonstrating potential mood-boosting properties of probiotics — were achieved using strains of bacteria in far higher amounts than are available in products on the market, scientists caution. Experts also warned me that dietary supplements are poorly regulated, with many lacking the extended shelf life required for anyone to reap health benefits from them. Even worse, they may not even contain the ingredients stated on their labels. So I'm reluctant to endorse any of those products, especially if they're unrefrigerated. But I'm comfortable making one dietary recommendation to harness gut bacteria for gain or, rather, loss: Eat more yogurt. It appears that the cultures in at least some commercial brands might help you maintain a healthy weight.

In one of the largest and longest epidemiological studies on the role of diet in weight gain, five experts on nutrition at Harvard tracked 120,877 health professionals — nurses, doctors, dentists, and veterinarians — for one to two decades. Every two years, the participants filled out detailed questionnaires about their diets and current body weights. Over each four-year period the average weight gain of individuals was 3.4 pounds, or 16.8 pounds over two decades. As might be expected, the foods associated with the widest girth in each four-year interval were such American staples as french fries, which were associated with a weight gain of 3.4 pounds; potato chips, 1.7 pounds; sugary beverages, 1 pound; and red meats, .95 pound. High consumption of vegetables, whole grains, fruits, and nuts, however, were actually associated with weight loss, in the range of .22 to .57 pound.

Yogurt topped the list of the most slimming foods, contributing to a .82-pound drop in weight over each four-year interval, or a loss of 4.1 pounds in twenty years. The lead investigator, Frank B. Hu, speculates that its bacteria cultures may stimulate the body to make hormones that reduce hunger, prompting high consumers to ingest fewer calories overall. Maybe all that yogurt in their bellies helped to lift their spirits too — something not tracked by the study but a plausible, if unproven, hypothesis.

NOW THAT WE'VE EXPLORED the many pathways by which gut bacteria can influence our moods and fundamental drives, let's turn to a question central to the theme of this book — namely, why did gut bacteria evolve to direct our behavior?

Here we are on shaky ground, but I offer this educated guess: We humans use our brains to create music, understand mathematics, and ponder the fate of the universe, so we tend to think of the gut as serving the brain rather than the other way around. But when bacteria began to colonize animals, some eight hundred million years ago, brains were not so sophisticated. Earthworms are believed to be among the first creatures to harbor gut bacteria, and each wriggling form basically consists of one long digestive tract surrounded by nerve fibers — what we now refer to as the second brain — to coordinate digestion. The main function of the brain in its head — if two tiny clumps of cells can be called that — is to obey orders from below like "Eat, eat, eat!" so as to keep the bacteria teeming inside its tube of a body well fed. According to microbiome expert Mark Lyte at Iowa State's College of Veterinary Medicine, the brain above may even have evolved as an outpost of the gut's nerve network, in which case the second brain came first. So from the start, gut bacteria have been in very close communication with the brain upstairs and maybe even quite dictatorial in their demands. After all, they greatly outnumber the rest of the cells in the body and clearly have a stake in the safety and welfare of their vessel. And as that vessel has evolved and its range of behavior grown more complex, its microbial inhabitants have, with the ever pressing need of nourishment, been compelled to extend their control beyond simple appetites to the realms of emotion and cognition. Which is why without them, suggests McMaster University's Stephen Collins, an animal can seem rudderless, listless, or downright reckless. As he points out, germfree mice don't learn very well or remember where they've been. They don't avoid predators. They don't become distressed or protest when separated from their mothers, whose nurturance and protection is critical for their survival. "But," said Collins, "if you colonize them with the normal microbiome for that strain, they calm down and behave in a much more appropriate, cautious manner. You could say that it's in the bacteria's best interest that the host survives and takes fewer risks."

UCLA's Mayer agrees: "These microbes have been cohabitating with us for a very long time, so there must be some benefits that they've offered in areas related to hunger, agitation, sexual behavior, aggression, and anxiety. These behaviors all evolved to enhance survival." Evolution, in his opinion, is not selecting for only the macroorganism or the microorganism — it's selecting for both. "It's optimizing the system. If you live in a dangerous environment," he hypothesized, "these bacteria don't attenuate but enhance fear since it's better to overreact to a threat in those circumstances. Or if you live in an environment of scarcity, the bacteria stimulate the dopamine [a neurotransmitter involved in reward] system so you're constantly searching at high risk for food."

HERE'S WHERE IT GETS COMPLICATED, however. Even though the fate of the superorganism and its constituent microorganisms are closely intertwined, their goals may not always be aligned. You may want to go surfing on a great big wave for the sheer thrill of it, but your gut microbes have nothing to gain from your taking that risk, and should you drown, they'll go down with the ship. To make matters still more confusing, these bacteria are themselves in competition with one another. One group might crave carbs and another high-protein fare, with each lobbying the host to act in its interest: *Have a bagel! No, no — go for the T-bone steak!* While this is a flight of fancy, it's possible, and even probable, the brain is indeed being bombarded by conflicting commands from gut microbes, though how such disputes would be settled is anybody's guess.

Whatever the mechanism of arbitration, these microbes seem, for the most part, to get on swimmingly with one another and us. They are not by nature an unruly, aggressive bunch intent on quickly killing their host and moving on. Compared to microbes selected for virulence, they live life at a more leisurely pace and are less intrusive, spreading much more slowly — in a bead of spittle when a mother kisses her child or by a handshake, especially if the owner of one of those hands forgot to wash it after going to the bathroom. They have traded a life of piracy and murder for a more settled existence — a roof over their heads, a warm meal they can count on. Still, like their hosts, they can be opportunistic. If they sense they can get away with it, they may gnaw their way into the stomach or inflict other kinds of harm. And they are at the mercy of monsters like rabies, microbes that have no intention of getting along with any living thing and attack the brain directly, manipulating the host far more effectively, to their and our loss (though when we die, gut microbes get to eat us). In short, gut bacteria are no more altruistic than parasitic manipulators — it's just that their survival strategy tends to be more closely allied with our own. And because they usually want us to act in ways that promote our well-being, they typically draw less attention to themselves than malicious microbes do. But make no mistake: Their influence on our behavior is dramatic. In fact, I'm not sure we will ever truly be able to separate their motives from our own.

Clearly much more research is needed to clarify the nature of this relationship, but so-called gut feelings almost certainly have a basis in human physiology. Psychiatry and gastroenterology may be more closely allied than we ever dreamed.

8

Healing Instinct

T HE CUNNING PLOYS AND BEHIND-THE-SCENES meddling of nature's puppeteers would make Machiavelli gasp in
awe. But hosts are no pushovers. In addition to having immune
systems and healthy gut bacteria to guard against infection, we and
other animals come into the world equipped with parasite-tracking
radar — an array of sensors attuned to the high-pitch drone of mosquitoes, foul odors, visible signs of disease, and far more subtle hints
that contaminants lurk nearby. When trouble is detected, this defense
system spurs us to take immediate evasive steps or, if we do fall ill, it
prompts us to react in stereotypical ways to reduce the damage. Scientists are discovering that these protective behaviors, which essentially
function like a shadow immune system, can be surprisingly sophisticated.

Creatures with no knowledge of the germ theory of disease have an
instinct for healing and staying well. Good hygiene, vaccination, therapeutic interventions — these are the pillars of modern medicine. Yet
animals of almost every stripe engage in these practices, as did early
humans. Indeed, were it not for these evolved defenses, the immune
system would quickly be overwhelmed.

One of the most familiar but widely misinterpreted examples of this phenomenon is what scientists call sickness behavior. When you're ill, you spike a fever, lose your appetite, and become depressed and listless. Contrary to popular belief, these symptoms don't mean that the disease agent is weakening you but just the opposite — they demonstrate that the brain, in conjunction with the immune system, is mounting a multipronged campaign against the invader. Infectious organisms typically can live only within a narrow temperature range, so fever kills them in droves basically by boiling them — a brilliant defensive tactic, but one that requires a vast amount of energy. To turn up the body's thermostat just one degree Celsius requires roughly the same number of calories that an average adult would expend walking forty kilometers. To funnel that much energy to the battlefield, the brain begins snapping orders: *Stop moving about! Stop looking for a mate! Stop foraging for food and spending precious energy digesting it! Stop everything and go to bed!* And so you fall into a feverish sleep.

Fever is so important for slaying germs that animals that cannot regulate their own body temperature — for example, locusts, baby rabbits, and cold-blooded creatures like lizards — have found alternative means to cook pathogens: They sunbathe. Lest anyone doubt that sickness behavior is a defense against pathogens, scientists can induce it without exposing animals to a single germ. They manage this simply by injecting healthy rodents with immune components called cytokines. The once-frisky animals refuse to eat or drink and lose their passion for running in their wheels. Up go their temperatures and down go their heads. They act and feel sick even though they're healthy.

Sometimes fever and related defenses are not enough to control rapidly multiplying germs and the slew of poisons they churn out. In such situations, the nervous system may aid the immune system by opening valves at key junctures in the GI tract while simultaneously reversing the rhythmic contractions of the intestine, forcing food back up from whence it came. Meanwhile, the brain begins signaling wave upon wave of nausea. You know what happens next: You throw up. Congratulations! In just a few convulsions, you've ejected an army of nasty germs.

Vomiting is not just a means of getting rid of harmful microbes but also a preventive measure. A phenomenon known as sympathetic vomiting occurs when the sight of someone throwing up causes others to do so too. Such copycat behavior probably evolved to protect us from food poisoning, a hazard that was both more common and more lethal in generations past. Imagine yourself seated beside a primeval fire, enjoying a communal meal of antelope stew, when the person beside you begins retching up his dinner. The stew may not have been to blame for his explosive bout of illness, but under the circumstances, it would be a wise precaution to react the same way, which is why nausea, scientists believe, can be highly contagious.

MANY SCIENTISTS HAVE ADVANCED our understanding of behavioral defenses against parasites, but Benjamin L. Hart, a veterinarian and neurobiologist at the University of California at Davis, stands out for the size and scope of his contribution. In the 1970s, Hart connected fever to other sickness behaviors, intuiting the evolutionary advantage of linking the traits. In addition to his own research, his overview articles have created a conceptual framework for the field, stoking scientific interest in the phenomenon. A man with an astute eye and a gift for seeing connections between disparate findings, Hart has not limited his investigations to the laboratory. Many of his insights stem from long sojourns that he and his wife, Lynette, a fellow faculty member in the UC Davis veterinary school, spent in Africa studying wild animals. When they first saw them up close, both recall being amazed by their robust health. "Being in a veterinary school, you know how much care we give our pets and zoo animals," he said. "We keep them in clean, sheltered environments, disinfect their wounds, vaccinate them, and give them antibiotics and other drugs. Wild animals, in contrast, get scratched and mauled, exposed to swarms of biting insects, and feast on carcasses dragged across dirt. Yet in spite of receiving no medical attention, their parasite levels are typically low.

They get along fine in nature without any intervention. Some of them thrive."

Before speaking with the Harts, I assumed animals — and the humans of yore — expended far more energy dodging big predators than tiny parasites. I now suspect the opposite is closer to the truth. And that doesn't take into account the cogs always spinning in the background of our minds, sparking inexplicable sexual attractions, a hunger for sleep, weird dietary cravings, and who knows what other peculiar urges — all to keep us safe from parasites. Meanwhile, we are oblivious to the origins of these drives or the minefield of hazards they help us navigate. Just warding off parasites that live on the surface of the body — lice, mites, ticks, mosquitoes, and the like — can be an exhausting, full-time job in itself. These real-life vampires come in endless varieties; humans and birds alone attract over two thousand species of fleas.

The term for these pests is *ectoparasites* (*ecto-* means "outside," signifying their tendency to live on rather than inside another animal's body) and they are not just a nuisance. Tabanid flies can siphon off a pint of a horse's blood in a day. Just half a dozen engorged ticks on a delicate gazelle or impala can weaken these fast sprinters, turning them into easy marks for predators. Warble flies can reduce the annual weight gain of cattle by twenty to seventy kilograms (about fifty to one hundred and fifty pounds).

Ectoparasites are all-around bad news; they sap energy, stunt growth, and diminish animals' success at winning territory and competing for mates. If the infested animals do manage to breed, they produce fewer offspring and less milk to feed them. Ectoparasites also carry microscopic parasites — the germs that cause scourges like malaria, dengue, and Lyme disease — which the insects transmit to their hosts as they feed.

Small wonder animals go to extreme lengths to avoid such pests. Rodents spend one-third of their waking lives grooming. Blue herons peck at the mosquitoes hovering beneath them an astonishing three thousand times an hour. Horses and other hoofed animals shake their heads, twitch their ears, stamp the ground, swish their tails, bunch together, and, if all else fails, gallop away. Some animals even make their own fly swatters. While in Nepal, Hart was surprised to see an Asian elephant break off a large branch of a tree, remove excess leaves to fashion it into a switch, and then shake it over its head and body. "It's tool use," said Hart, "and their switch really does cut down on flies." Gazelles and impalas take a different approach to dislodging parasites: they use their teeth as combs, scraping ticks off their coats and orally grooming themselves two thousand times a day.

These heroic efforts are not for naught. Mice prevented from grooming have sixty times more lice. Those herons fiendishly pecking at mosquitoes stop 80 percent of their would-be attackers from robbing their blood. And in a study the Harts conducted on wild impalas, they showed that grooming behavior reduces an animal's tick load twentyfold.

Pest-avoidance strategies can be more sophisticated than meets the eye. Gazelles and impalas, the Harts discovered, groom even when they have no ticks. Several times an hour, said Benjamin Hart, "a clock in the animal's head" tells it to systematically scrape clean the very path that ticks typically travel en route to the head or rear end, where they're out of the animal's reach. The phenomenon, which the couple dubbed programmed grooming, is now well accepted, but they initially had a hard time selling their peers on the idea, he recalled. "They'd say, 'The animal itches. They groom because they itch.' I'd say, 'Ticks are very good at not being felt when they bite.' Their response: 'Oh, Ben, why are you wasting your time studying that?'" Finally, after returning from Africa, the couple went to the San Diego Zoo Safari Park and recorded the behavior of Thomson's gazelles, which were tick-free in their sanitary enclosure. Yet they groomed themselves anyway, just like clockwork.

The Harts' findings may have initially been greeted with skepticism because scientists for the longest time underestimated the damage ectoparasites could inflict on their hosts. As Lynette Hart explained, "They thought grooming felt good and was all about social bonding."

Of course, for many gregarious species, it is about bonding too — but it's also nice to have another set of hands, paws, or teeth for removing hard-to-reach parasites. Mice, penguins, deer, and primates, among other species, maintain lower pest levels with the help of a mate or buddy than they would if left to their own devices. In impalas, Lynette reported, reciprocal grooming is done strictly tit for tat and is another example of programmed behavior — they do it whether or not either party has ticks. One impala walks straight up to another and delivers a bout of eight or so grooming strokes to the neck. If its partner does not immediately reciprocate, and virtually the same number of times, the impala will go somewhere else. "They don't allow much cheating," said Lynette. Wood mice and long-tailed macaques have adopted a different barter system: these species pay for grooming with sex.

Other animals outsource the job to different species. This is an especially common arrangement in the marine ecosystem. Large fish arrive at special territories called cleaning stations and pop open their mouths; in swim little fish and tiny shrimp, which eat up all the parasites and other tidbits clinging to their teeth and gills. Oxpecker birds perform a similar function in Africa for rhinos, buffalo, zebras, giraffes, and eland. One of the bird's specialties, said Benjamin Hart, is cleaning out ears — a favorite hiding spot for ticks. "You can see animals making these little postural shifts to allow the birds to get deep in there and dig them out."

WILD ANIMALS' RESILIENCE is all the more remarkable when you consider that the vast majority of parasites are too small to be detected with the unaided eye. Given that nature's creatures don't come into the world equipped with microscopes or medicine cabinets, how are they so successful at warding off infection?

One of their many tricks is to avail themselves of what Hart calls "the medicine cabinet in their mouth." When injured by a bite, gash, or scrape, numerous species — among them primates, felines, canines, and rodents — "use their tongue like an antiseptic wipe to clean wounds." Saliva is rich in antimicrobial agents, immune-boosting substances, fungicides, and growth factors for stimulating healing of both skin and nerves. In laboratory experiments, the removal of rodents' salivary glands retarded the healing of their skin wounds. In another study, a sheet of human cells grown in culture was punctured to simulate a wound. The addition of saliva to the petri dish prompted cells in the area of the injury to grow much more quickly than those not so treated. In the right situation, said Hart, "the old saying 'go lick your wounds' is very good advice."

Like primates today, our ancestors probably licked their wounds too, he and Lynette believe. Modern humans may continue the tradition, albeit perhaps unknowingly. A few days after my conversation with the Harts, I accidentally nicked my finger while slicing an orange. Instantly I began sucking on the cut. Only after my finger had flown into my mouth did I recall their words or think to reach for soap and water.

Saliva can prevent germs from entering the body through other routes. After copulating, male rodents, cats, and dogs will furiously lick their penises for several minutes. The liberal application of saliva kills several pathogens that are leading causes of STDs in these species. Their habit also benefits females because it prevents males from passing on infections to their next mates.

Interestingly, cattle and horses, which can't lick their own penises, are much more prone to STDs — one of the reasons, according to the Harts, they're bred by artificial insemination. Humans, they add, are also very susceptible to STDs, possibly owing to similar anatomical limitations.

Lactating females of many mammalian species have found still another healthful purpose for saliva. They use their tongues to sponge away germs on their nipples before allowing their young to nurse. "Rodent pups may even refuse to attach to her tit unless it has first been washed with her saliva," said Hart.

Another smart move for staying free of parasites is to avoid those mounds of germs we call feces. Humans recoil at the sight and smell of bodily waste — and it serves us well. Contact with fecal contaminants can expose us to a long list of hazards, including numerous varieties of intestinal worms, cholera, typhoid fever, hepatitis, and rotavirus (a leading killer in the developing world).

Poop poses just as broad an array of dangers to other species, many of which react to it much as we do. Chimpanzees have — in Jane Goodall's words — "an almost instinctive horror of being soiled with excrement." When they accidentally come in contact with feces, they grab fistfuls of leaves and vigorously wipe it off. Even sex can quickly lose its appeal if excrement enters into the picture. Goodall reported that when a female chimp signaled her eagerness to copulate by raising her rump to a male, he initially seemed game until he spotted a diarrhea smear on her fur, whereupon he opted to abstain. Another male, no doubt with fewer prospects, eventually took her up on her offer, but not before first meticulously wiping off the offending spot with leaves.

Other animals are equally fastidious when it comes to feces. Mole rats and other small mammals that live inside burrows build underground latrines that are separated from their living quarters and larders. Lemurs in Madagascar have their own version of an outhouse — mounds aboveground, which they visit only to relieve themselves. Cows, sheep, and horses don't graze near fresh dung heaps no matter how lush the grass is in the vicinity of those piles.

Wolves, hyenas, and big cats never soil their own dens — an instinct, said Hart, that accounts for why humans find it so easy to housetrain their domesticated relatives (that's not true for all dog breeds, however! Shih tzus and miniature terriers can take years to housetrain, he cautioned, because that instinct was diluted by their human breeders). Fish, too, have their taboos about where it's not proper to poop. When nature calls, several species are known to swim to the edge of their home range or just beyond.

Even the birds and the bees have good potty habits. Northern flicker woodpeckers remove their nestlings' poop — which conveniently comes wrapped in gelatinous sacs, the avian equivalent of a diaper — fifty to eighty times a day (for comparison, a human baby gets roughly that number of diaper changes in a week). As for bees, some are known to make a bathroom run a group event. When they feel the urge to go, they fly away from the hive and relieve themselves all at once, showering a yucky yellow mist on whoever is below. On a visit to Laos in 1985, Alexander M. Haig Jr., then secretary of state, mistook clouds of bee poop for chemical warfare.

Avoiding the sick is no less important for staying well. Down through the ages, humans have been quick to flee and isolate those they feared could spread much-dreaded diseases like leprosy, bubonic plague, TB, polio, virulent strains of the flu, and, later, AIDS and Ebola. Reasoning — albeit highly primitive over most of history — and advancing medical knowledge have motivated these actions. But there may be an innate component to these impulses too, for we are not alone among animals in shunning those who display signs of disease.

Bullfrog tadpoles can sense a chemical signal in the water emanating from a tadpole with a yeast infection of the GI tract, whereupon they swim in the opposite direction. Likewise, rodents can detect the odor of parasite-infested members of their own kind, whom they then avoid or keep at a distance through aggressive displays. Killifish treated with a dye to imitate the dark splotches caused by a trematode infection are not as attractive shoal mates as their unmarked counterparts. Apes show similar tendencies. A male chimp who was never short of grooming partners when healthy, Goodall reported, became an outcast after polio cost him the control of his hind limbs and flies began to hover around him.

SHUNNING SOURCES OF CONTAGION is an excellent first line of defense, but boosting the body's resistance to infection is equally critical. We think of vaccines as a sophisticated tool of modern medicine, but animals and humans without MDs have also figured out ways to vaccinate. When an ant gets a deadly fungus, for example, another member of its colony will rush up and lick it, thereby exposing the insect to a tiny dose of the disease agent. This method of inoculation is not without perils — 2 percent of ants perish. But the vast majority develop heightened immunity to the infection, according to evolutionary biologist Sylvia Cremer, who studies the insects.

Interestingly, their approach resembles a technique for vaccinating against smallpox that was used by people in northern Africa long before an eighteenth-century English physician named Edward Jenner received credit for inventing the world's first vaccine against the disease. In this ancient practice, a scab from someone with smallpox was rubbed into a tiny cut on the skin of a healthy person. And just as in the case of inoculated ants, 2 percent of those people died. The custom, however, appears to have averted far more tragedies than it caused, lowering the mortality rate from smallpox by 25 percent. (Today's vaccines, it should be noted, contain only *non*infectious components of pathogens, and the risk of death from their use is infinitesimal.)

Mothers of newly weaned carnivores like lions may take a messier approach to vaccination, reported Hart. They drag the cubs' first serving of red meat across the den floor, coating it in bacteria and dirt. These contaminants make for a healthy meal because they prime the young animals' immune systems to better withstand the barrage of germs they'll soon encounter when they begin hunting themselves. Indeed, claylike compounds similar to those found in dirt are often intentionally put into medical vaccines today because such adjuvants, as doctors call them, enhance the immune system's response.

Some species of monkeys are thought to inoculate their young by passing their newborns from one member of the colony to the next. Exposing them to communal germs so soon after birth may seem like a bad idea, but they must be able to fend for themselves early on, said Hart, "so their immune systems need to mature as fast as they do."

Primates that develop more slowly, like humans, can afford to insulate newborns from all but the closest of kin until, little by little, their defenses strengthen. As soon as babies start to crawl around, however, they seem programmed to put virtually any object within reach into their mouths, including, inevitably, all sorts of cringe-worthy things — the sponge used to clean the sink, a pacifier dragged across the floor, what the cat spit up, rocks and snails from the garden, and, of course, dirt. Babies normally swallow as much as a gram each day.

Such oral inclinations have traditionally been explained as exploratory activity — how babies learn about the tastes, textures, and other properties of objects around them. But nature is famous for multitasking, so it may well be that sampling their surroundings is training both their senses *and* defenses. Maybe their adventurous palates are one more example of animal-style immunization.

Complementing that view, proponents of a popular, though still hotly contested, theory — the hygiene hypothesis — believe that rearing children in too clean an environment may make them more susceptible to allergies and other diseases. These scientists believe that an immune system that is repeatedly tested early in life becomes sharper at distinguishing between "good" and "bad" germs, making it better at gauging when to mount a counterattack. Recall also that microbiome research has begun to link more diverse gut bacteria to better health, and those microbial populations are established early in childhood. So it certainly makes sense that babies have evolved to increase their uptake of germs once the dangerous period early in infancy has passed.

All these ideas need stronger confirmation. But if they are proven to be valid, they have provocative implications; it may be that in our zeal to create safe, sterile environments for our children, we have replaced natural inoculation with manufactured vaccines, antibiotics, and allergy medications. No one would want to return to the era before these advances, when many babies never made it to adulthood. But perhaps a little laxity in our housekeeping and hygiene standards might do more good than harm. Pharmaceutical corporations might even take a lesson from nature. Imagine pills of the future that contain thousands of the bacteria in soil that help stimulate immunity but none of the nasty germs like toxoplasma and toxocara — basically, a safe-to-consume mud pie.

WE'VE FOCUSED ON BEHAVIORS that might have evolved to reduce our susceptibility to pathogens. But there is one very risky activity that's hard to avoid and so must be undertaken with extreme care. That activity is sex. The commingling of bodies during mating offers parasites a wonderful opportunity to jump hosts and sicken more victims. Should those pathogens invade the genital tract, moreover, they can cause sterility; if a female becomes impregnated by an infected partner, her offspring might be born deformed or diseased. Any trace of illness in a prospective mate might signify that he or she has a weak immune system, a deficiency that, if passed down to the next generation, could doom one's lineage. To avoid these steep costs, animals have evolved courtship rituals that typically involve displays of great strength and vigor. And the least hint of illness or weakness in the animal kingdom is invariably a sexual turnoff.

Like men who pump iron to impress women with their bulging muscles, male guppies court females by repeatedly flexing their bodies into an S shape. Fish infested with parasites perform fewer of these displays, with predictable results: the weaker suitor is the loser. Female mice are just as choosy. When it comes time for coupling, they'll sniff a prospective partner, and if his smell betrays the presence of a pathogenic protozoan in his GI tract, he will be snubbed. As part of their mating ritual, sage grouse puff up yellow air sacs normally hidden under their breast feathers. Should an inflated neck reveal the swollen bite marks of lice — or red spots applied by a scientist to mimic them — a female will look for another partner. In many other bird species, females shun prospective mates with lackluster plumage and ornamentation — another indicator of parasitic infection. In one telling experiment, red jungle fowl were infected with an intestinal nematode. Compared to uninfected males, they had duller eyes and combs, shorter tail feathers, and paler hackle feathers. Underwhelmed by the drabber males, the females were half as likely to mate with them than with healthy males.

Humans may also rely on visual cues to find partners with robust immune systems — an indication that they're likely to endow heirs with heightened resistance to germs. Beauty is associated with not just youth and fertility but also health and previous successes in fighting off pathogens. Good looks, for example, are denoted by symmetrical features — a sign that early life development was not disrupted by infection — and skin that shows no trace of pockmarks, sores, or other blemishes. With that in mind, you'd expect beauty to be more valued by those more susceptible to germs — a theory that evolutionary biologists put to the test in a survey of over seventy-one hundred people on six continents. In keeping with their prediction, those who lived in countries where parasites were leading causes of death and disability — in Nigeria and Brazil, for example — deemed good looks much more important in a mate than did inhabitants of nations like Finland and the Netherlands, which have among the lowest incidences of infection. In a British study, merely prompting people to think of germs — by, for example, showing them photos of a festering skin sore or a white cloth with a dark stain resembling a fecal smear — boosted how much they preferred symmetrical faces in the opposite sex.

Not just our eyes but also our noses may guide us in choosing partners whose genes will endow our children with heightened protection against infectious disease. That claim — which is still contentious — has gained traction as a result of scientists' deepening understanding of a large cluster of two hundred or so genes called the major histocompatibility complex, or MHC. These genes code for cell-surface molecules that enable the body to distinguish its own cells from foreign invaders, marking the latter for destruction. Most people are familiar with MHC genes in the context of organ transplantation. The success of such an operation depends on the donor and recipient sharing many of those genes in common; a mismatch could spark an immune attack, leading the organ to be rejected.

In the 1990s, experiments on rodents revealed that MHC genes do more than just control a mouse's ability to detect foreign cells. They also determine its characteristic body odor. Put more simply, scent broadcasts vital information about the inner workings of an individual's immune system to other animals. What's more, the rodents preferred to mate with animals whose MHC genes were the least like their own, a choice based on a prospective partner's scent profile.

Their progeny, as a result, had more diverse immune genes, and that variability enhanced their survival. When parents have similar immune genes, their offspring tend to be susceptible to the same germs, so if one animal in a litter is killed by an infection, the rest are likely to perish with it, spelling the end of a family lineage. That liability is greatly amplified if, owing to a drastic decline in a group's population size, its members become inbred. Under those circumstances, a single viral infection can decimate the group's numbers. Inbred populations are also at greater risk of suffering from hereditary disease as a result of individuals inheriting two recessive genes — one from each parent — for a deleterious trait. Incest — an extreme form of inbreeding — carries the same risks, only further elevated.

In the case of humans, knowledge of the harmful physical and psychological consequences of incest no doubt contributes to the prevalence of this taboo in most societies, but it's also a rarity in nature. Indeed, incest accounts for fewer than 2 percent of births in wild-animal populations. Scent, scientists began to think, might be nature's way of discouraging unfavorable biological pairings.

Inspired by the animal research, a Swiss zoologist, Claus Wedekind, sought to determine if smell might have an unconscious influence on our species' mate preferences. In a famous experiment conducted on college students at the University of Bern, he gave males clean cotton T-shirts and instructed them to wear them for two nights and avoid using deodorant or any other scented products over that period. After collecting the sweaty shirts, he asked women to sniff them and rank their odor according to which they liked best. Dovetailing with animal studies, the women preferred the smell of men whose MHC genes were the most dissimilar from their own. And the male odors they liked the best reminded them of the scent of a current or previous sexual partner. More recently, Wedekind and colleague Manfred Milinsky made another intriguing discovery: People who share certain portions of MHC genes tend to prefer the same perfume scents. Perhaps, they speculate, we unconsciously choose fragrances that enhance our natural body odors, explaining the widespread use of perfumes over human history.

Meshing nicely with these results, independent studies of Europeans as well as Americans of European descent — notably the Hutterites and Mormons — suggest MHC opposites are indeed drawn to each other. In a related line of research, investigators theorized that where pathogens are most prevalent, natural selection should favor the survival of people with more diverse MHC genes. After analyzing human genome data collected from sixty-one populations around the globe, they found exactly that. Also complementing Wedekind's work, research conducted in the laboratory of Rachel S. Herz, an expert on the psychology of smell, at Brown University showed that women rate body odor as the single most alluring (or off-putting!) physical trait in a man. If a man's natural odor smells "wrong" to a woman, Herz's research suggests, she won't want to have sex with him no matter how dazzling his other qualities. Herz did not look at her subjects' MHC genes, but she strongly suspects a woman's preference for some scents over others is likely shaped by whether a man's odor signifies his immune-system genes will complement her own.

Not all the evidence ties together so neatly, however. Studies of the Yoruba of Nigeria and Amerindians of South America failed to find that mate choice was influenced by a partner's MHC genes. Other research suggests MHC odor preferences do influence whom a person finds sexually attractive, but in a more complicated fashion than previously assumed. Among other things, it appears that such preferences can be altered by early life experiences.

Strong support for this view comes from an experiment in which female mice were separated at birth from their own litters and placed in the litters of other mothers. Upon maturing, the females would not copulate with the unrelated males who nursed beside them on the same mother's nipples — their "adopted" brothers, if you will. In a reversal of the natural order, they chose to mate with their own genetic brothers. This suggests the scents of those who surround us when we're young leave a lasting imprint on the brain, defining our sense of kin and the kinds of odors we will later find sexually enticing. Nature's rule of thumb might be summed up thus: If he or she smells like someone you grew up with, look elsewhere for a sexual partner.

Of course, no comparable experiment has been performed on humans, but circumstantial evidence hints that a similar form of olfactory imprinting may occur in our species. It's long been known that unrelated children who are raised together in a communal environment such as a kibbutz almost never marry each other, perhaps because their scents — familiar since infancy — falsely mark them as biological siblings. The law has one definition of incest; the brain may have another.

Eons ago — hundreds of millions of years before humans showed up on the scene — sex itself may have evolved as a defense against parasites. To understand why, consider the alternative: asexual reproduction, also known as cloning. A strategy favored by single-celled organisms like bacteria, it entails simply splitting in two to create identical copies of the parent. Because clones multiply at a ferocious pace, they rapidly accrue beneficial mutations that allow them to overcome their host's defenses. Multicellular hosts like ourselves replicate too slowly to rely on new mutations to protect us. We need a better shield.

The invention of sex is evolution's solution to that challenge, according to a popular theory championed in the 1990s by evolutionary biologists Leigh Van Valen and William D. Hamilton. After briefly falling out of favor, the theory now appears to be ascendant again, buoyed by recent studies of animals both in the wild and in laboratory settings. What makes sex such a game changer, its proponents believe, is the simple fact that genes get reshuffled every time they're passed down from parents to their offspring. If you're human, that includes the two hundred or so genes that make up the MHC, which can be inherited in billions of different combinations. Thanks to sex, each of us is biologically unique, which means you and your neighbors will vary in your susceptibility to pathogens. When a deadly bug comes to town, you won't all die. Even as Ebola — an unusually vicious virus — raged through West Africa at the end of 2014, roughly 30 percent of infected people managed to survive its ravages, and this despite receiving minimal or no medical care. Many of these lucky people may have lacked a molecule on the surface of their cells that the virus must latch onto in order to invade them.

Precisely because we aren't clones, we wear different targets on our backs — an array of cell-surface molecules that serve as docking sites for different pathogens — and the prevalence of these markers within a population can fluctuate dramatically depending on the latest infection going around. No sooner has a pathogen locked on to one target than that docking site (and the people who carry it) diminishes in frequency due to the success of the microbe's own killing spree. The germ's destructive power declines until it mutates into a new virulent strain that can zero in on a different target. In subsequent generations, meanwhile, people sporting the first target may resurge in numbers because the rapidly evolving pathogen is now pursuing other quarry. In this way, Van Valen and Hamilton postulated, we and other sexual reproducers stay one step ahead of the parasites nipping at our heels — and so we chase each other round and round. (This theory is called the Red Queen hypothesis after the character in Lewis Carroll's *Through the Looking Glass* who says to Alice, "It takes all the running you can do, to keep in the same place.")

A major appeal of this hypothesis is that it would explain a profound mystery — namely, how a mode of replication as slow and inefficient as sex, which requires considerable exertion, comical contortions, and a willing mate, can persist side by side with cloning. Just think: Bacteria can easily churn out more generations of offspring in a day than humans produce in a millennium. Yet sex enables us to compete in the same arena.

SLEEP IS A PUZZLE that has occupied many great minds, for, just like sex, its evolutionary benefits are far from obvious. When deep in slumber, we are highly vulnerable to predators, and the extended blackouts cut down on time spent foraging for food, searching for mates, and caring for children. So why spend so many hours in this dangerous and unproductive manner?

A novel theory posits that sleep evolved to shunt resources that would normally sustain waking activities toward the immune system. Recall that people sleep more when fighting an infection to meet the defense system's increased demand for fuel. In times of peace, the army still needs to be fed and the troops replenished. Indeed, immune cells gobble up nutrients at a ferocious pace and have rapid turnover rates. One major type, the granulocyte, is so short-lived that these cells must be replaced every two to three days. Even when we're healthy, according to this view, a daily period of inactivity is essential for covering the steep energy costs involved in keeping the war machine ready for battle.

The immune theory of sleep, as it's been dubbed, aligns with the wisdom of mothers, who have long been telling their children to get a good night's rest, intuitively grasping sleep's necessity for staying well. The theory is also gaining support from several lines of research.

Sleep deprivation, animal studies show, does indeed increase susceptibility to infection. Also consistent with that observation, insufficient sleep immediately before or after receiving a vaccine reduces the body's immune response by half, greatly diminishing the protective benefits of inoculation — something to keep in mind when you're due for your next flu shot. Perhaps the most persuasive evidence that sleep boosts our resistance to parasites comes from research conducted by an international team of evolutionary biologists led by Charles Nunn and Brian Preston. The group gathered data on the sleep habits of twenty-six species of mammals with daily hours of slumber ranging from a skimpy 3.8 in sheep to a luxurious 17.6 in hedgehogs. The scientists found that across these species, more sleep strongly correlated with enhanced immune function as measured by the number of immune cells circulating in their blood. What's more, the longer a species typically dozes, the lower its level of parasitic infection.

Traditional theories of sleep underscore its importance in consolidating memory and learning, as well as its role in removing waste products from the brain. As there's strong support for these ideas too, it appears that sleep may well rejuvenate both the brain and the immune system.

That means skimping on it, as so many of us do, may be inviting double trouble.

ANIMALS NOT ONLY ACT in stereotypical ways to reduce their risk of infection; the more resourceful among them also exploit compounds in their habitat to that end. Even more impressive, some have a knack for choosing medicinal plants. Intriguingly, reports Benjamin Hart, their criteria are the same as traditional healers' — both groups have a tendency to seek bitter cures for infection. After gathering data on the palatability of twenty-two herbal remedies, he found that sixteen were bitter — and some healers specifically equated the potency of treatments with their bitterness. This belief has a basis in science. Bitterness is a measure of toxicity, so these compounds are often effective in killing germs, intestinal worms, and other parasites.

The expression *a bitter pill to swallow* recognizes the unfortunate truth that powerful cures are often foul-tasting. Mary Poppins recommended a spoonful of sugar to help the medicine go down. The modern pharmaceutical industry has tried to disguise medications' unpleasantness with grape and cherry flavors. Yet since antiquity, humans have endured these awful remedies — but only when sick. Otherwise we spit them out.

Some species of tiger moth caterpillars also reverse their food preferences when ill. Normally they won't eat the bitter plant *Plantago insularis*, but when parasitized by insects, they suddenly develop a craving for it. As it turns out, the poisons in the plant's leaves are toxic to the parasites.

Wild chimps show exactly the same tendency. When suffering from diarrhea and other patent signs of infection, reported primatologist Michael A. Huffman, they'll take long detours from their normal route in search of *Vernonia* — a toxic, astringent-tasting plant with compounds thought to inhibit the growth of amoebas, bacterial pathogens, and intestinal worms. Upon finding it, an ill animal will meticulously peel away the bark of a fresh shoot, exposing its pith, and suck on its very bitter juices. Healthy bystanders in its troop may watch as it slurps up the juices but don't themselves partake. The young and curious, however, sometimes need discouragement, Huffman observed. On seeing a sick adult toss away the remnant of a *Vernonia* pith, a baby chimp attempted to pick it up, evidently with the intent of tasting it. His mother stepped on the pith and carted him away.

Goodall recorded a similar tale: With the hope of coaxing ill chimps to take a bitter antibiotic, she tucked the remedy into bananas. The sick animals readily ate the fruits, medicine and all, but healthy members of the colony would eat only unadulterated bananas.

Herbivores are known to consume a wide variety of leaves, berries, fruits, and other plant components with medicinal properties. Not all of them are bitter and toxic — or, if they are, only mildly so — and consequently, they may be part of an animal's everyday diet. *Aframomum,* a plant in the wild ginger family with a pungent, spicy flavor, is a case in point. It makes up a large proportion of the foods consumed by western lowland gorillas, who eat its pith and seedpods. Phytochemist John Berry has shown that these parts of the plant are very high in antimicrobial compounds that act in a manner akin to narrow-spectrum antibiotics; that is, they don't kill bacteria indiscriminately but rather thwart pathogenic strains such as salmonella and shigella while encouraging the growth of healthy gut bacteria.

People in some villages in West Africa also eat *Aframomum.* It's customary to serve the seedpods — called grains of paradise — when guests come to visit, much as North Americans might set out a bowl of peanuts as a snack. In addition, the plant is viewed as a medicine. In Uganda, people take *Aframomum* for infections caused by bacteria, fungi, and intestinal worms.

The distinction between food and medicine can be blurry. That's particularly true of a culinary tradition most of us take for granted — adding a dash of this or that spice to our dishes. Many of these seasonings double as antimicrobial agents, according to research by Cornell evolutionary ecologist Paul Sherman and his former graduate student Jennifer Billing. Before the invention of refrigeration, that benefit would have been of immense value — likely the reason, they believe, that monarchs sent legions of men to fight wars, cross oceans, and explore new lands in order to procure spices. Of thirty spices for which the scientists were able to collect data, all killed at least a quarter of bacteria species. Half of the spices inhibited the growth of 75 percent of bacteria. Some of the most widely used — garlic, onion, allspice, and oregano — killed every species they were tested against.

Oddly, they found that pepper had only weak antimicrobial effects. How to explain its immense popularity? Their research revealed that it acts synergistically with other spices, dramatically amplifying their bacteria-killing power. An example is *quatre épices* — pepper, cloves, ginger, and nutmeg; these spices are used together so often in France, especially for the preparation of sausages, that the mixture goes by its own name. Sausages are an excellent growth substrate for *Clostridium botulinum* — the cause of deadly botulism — so seasoning them with any old spices might be ill-advised. Other famous mixes such as curry powder (made from twenty-two spices) and chili powder (made from ten) similarly turn out to be, in the words of the scientists, "broad-spectrum antimicrobial mélanges."

Since food, especially meat, spoils more quickly in hot climates, Sherman and Billing predicted that the highest use of spices in the past would be in tropical regions — an idea they tested by combing through the ingredients of thousands of meat-based recipes from centuries-old cookbooks. True to their expectations, in warmer countries — notably Thailand, India, Greece, and Nigeria — all traditional dishes with meat used at least one spice and some twelve or more. Our ancestors from colder climates, in contrast, had much blander diets. For example, one-third of traditional meat-based recipes from Scandinavia called for no spices at all. A subsequent study of traditional recipes containing only vegetables — a less inviting growth medium for bacteria — found that they had fewer spices than those with meat, providing further support for their hypothesis.

Why anyone thought to use spices in the first place is more challenging to explain. After all, they have little or no nutritional value and are not very appetizing on their own.

Since even animals are drawn to bitter substances when ill, one likely scenario is that spices were first consumed as medicines. Indeed, garlic, turmeric, ginger, and cumin, to mention just a few, figure prominently in ancient folk cures. In small doses — mixed with meals — spices become more palatable. In other words, consuming them with food (think of Mary Poppins's "spoonful of sugar") improved patient compliance.

Whatever originally prompted early humans to ingest spices, numerous factors would have favored their incorporation into everyday diets. As Sherman and Billing pointed out, the custom not only cut down on food-borne illness but also allowed people to preserve food longer — a huge advantage in times of scarcity. In addition, dishes with spices in them may have tasted better, at least to some palates, promoting their broader acceptance. Families who used spices may have touted their many benefits to their neighbors. Another mode of cultural transmission would have occurred in the womb. In the last trimester of pregnancy, we now know, babies sample their mothers' diets by ingesting amniotic fluid. After birth, they show a preference for flavors that they encountered prenatally, biasing them toward eating foods and spices that are healthful in their communities. A passion for spices may even have been genetically transmitted, suggested Sherman and Billing. Early on, spice lovers — especially in hot climates — would presumably have left more heirs than those who couldn't stomach them. "Darwinian gastronomy," they pithily remarked in an article in *Scientific American,* may explain "why some like it hot."

Adding spices to meals may be an early example of preventive medicine, but a few other culinary practices for cutting down the risk of food poisoning are at least as old, if not far more ancient. Familiar to most of us, these include the salting, smoking, and, most important, roasting of meat. Indeed, the harnessing of fire for cooking, first clearly documented at campsites occupied by Neanderthals five hundred thousand years ago, remains the single most widely used weapon for zapping germs that lurk in meals.

ANIMALS THAT LACK OUR ABILITY to preserve and cook food must of course adopt other behaviors to prevent food-borne infections. For dogs and cats, one such behavior is eating grass. As Benjamin Hart explains, they do so "to flush intestinal worms out of their systems. The animals usually can't know if they have intestinal worms so they occasionally eat grass as a form of prophylaxis." Puppies and kittens do this most often because their small size makes them particularly vulnerable to the energy-draining effects of parasites. Our pets inherited this behavior from their wild ancestors. Wolves and cougars, for example, regularly eat grass, which is found in about 2 to 4 percent of their scat, sometimes along with worms they've expelled. The consumption of leaves, according to primatologist Huffman, may serve the same function in chimps, bonobos, and lowland gorillas. The leaves they choose always are covered in indigestible hairs, and the animals never chew them, as they would food, but rather swallow the leaves whole — sometimes as many as a hundred at a time. All that roughage, Huffman believes, dramatically accelerates the movement of food through the GI tract, purging them of at least two species of parasitic worms.

Animals can't call exterminators when their homes become infested with pests, but fortunately some have found methods of handling such problems — ones, in fact, that are not all that different from our own. Some birds and rodents — especially species that reuse the same nests generation after generation — drive unwanted guests away with toxic vapors, or so think many biologists, who term the strategy *fumigation*. During breeding season, such creatures can often be seen sprucing up the interiors of their homes with fresh leaves that they weave into old bedding or the thatching of last season's nest. "They don't grab just any green twigs that happen to be nearby," said Hart. "They hunt for leaves that are strongly aromatic and rich in volatile chemicals. These are markers that they'll be good insecticides, fungicides, and antimicrobials." One plant favored by birds for this purpose is fleabane, so named by ancient herbalists in recognition of its flea-repellent properties. The dusky-footed wood rat prefers California bay laurel, whose leaves they nibble on to release their poisonous fumes. Entomologists use bay laurel in insect killing jars. They also report that the specimens are more resistant to mold when subsequently pinned to display boards — a clue that rodents may use bay leaves as a fungicide as well. In field experiments involving birds, removal of fresh green twigs from spruced-up homes had predictable results; one starling nest, for example, became overrun with mites.

Nature's pharmacopoeia contains a bounty of insecticidal and antimicrobial substances that animals harness for their healthful properties. Wild Kodiak and brown bears dig up osha roots (*Ligusticum wallichii* and *L. porteri*), chew on them to release their volatile oils, and then work the paste deep into their pelts. Hinting at the root's medicinal value, the Navajo people use it as an antibacterial and anesthetic salve, and according to their legend, the burly beasts taught them about its healing power.

In Panama, white-nosed coati, cousins of the raccoon, travel long distances to obtain the menthol-smelling resin of *Trattinnickia aspera* trees, which they vigorously smear over their bodies with their paws. Compounds in the resin, chemical tests show, repel fleas, lice, ticks, and mosquitoes.

Not all the drugs animals reach for are botanicals. Wedge-capped capuchin monkeys in Venezuela roll their bodies over millipedes to stimulate the insects to release defensive toxins and then frenetically apply the chemicals to their fur with a liberal quantity of their own drool. The millipedes evolved the toxins to repel their insect enemies, so the monkeys are, in effect, stealing their bug spray.

Birds — some two hundred species of them — use a similar strategy. They crush ants with their beaks, causing them to release their own version of insect repellent, then rub the bugs through their feathers.

Right underneath our feet is another much-prized natural medicine. It shields the gut from a long list of food- and waterborne pathogens, properties that are sought after by droves of animals and hundreds of thousands of people around the world. This potent elixir is dirt.

Both humans and animals are very picky about what dirt they'll eat. They turn their noses up at the dark, coarse topsoil that babies ingest when exploring their surroundings. The earth craved for its medicinal value is typically collected from ten to thirty inches beneath the ground. Often light in color, it has a high content of clay that is chemically similar or identical to kaolin, the active ingredient in the original formulation of Kaopectate — the world's best-selling treatment for diarrhea and nausea. Owing to the molecular structure of such clays, they clasp hold of viruses, bacteria, and fungi, all of which then get flushed out in our waste. These compounds similarly bind to poisons produced by pathogens as well as toxic chemicals found in plants that are dietary staples. To add to the benefits of medicinal clays, their fine, slippery texture coats the mucosal lining of the gut, reinforcing the body's natural barrier against parasitic invaders, including amoebas, roundworms, and flatworms.

Extensive field observations of five wild chimps at Mahale Mountains National Park in Tanzania show that the animals will grab a handful of clay from termite mounds when suffering from diarrhea and other gastric ailments. According to Cindy Engel, a biologist and author of the book *Wild Health,* herbivores like gorillas, elephants, rhinos, and parrots obtain the clay from a variety of sources, such as eroding riverbanks, powdery volcanic rocks, and large patches on the ground that they've denuded of vegetation. The clay allows them to extract nutrients from plants that would otherwise be too toxic, but Engel believes these animals also benefit from its power to purge parasites from their systems. Clay is especially valuable to rats, she reported, as those animals can't vomit. When poisoned with lithium chloride by experimenters, they immediately eat clay if given the chance and are presumed to do the same when their gastric upset is caused by pathogens.

People have been consuming clay since at least as far back as ancient Greece and Rome. The practice, called geophagy, continues to this day in traditional societies on every inhabitable continent. Australian Aborigines get their medicinal dirt from termite mounds. In sub-Saharan Africa, earth from clay pits is baked in the sun, air-dried or heated, and then sold in markets for digestive ailments and morning sickness. Slaves brought the custom to the rural American South, where it had already been embraced by Native Americans and later was adopted by poor whites disparagingly called "sandlappers," "sand-hillers," and "hillbillies." Though the practice in recent years has been condemned as disgusting and deviant, geophagy is hardly moribund in the United States. Many geophagists report that they simply can't resist the richness of earth, its delectable aroma, and the way it "melts like chocolate" on the tongue. "When I'm pregnant," one American aficionada of dirt told *Time* magazine, "it's just like taking dope."

Highlighting clay's importance in combating infection, a global survey shows that geophagy is most popular in the parasitic hot zone of the tropics, steeply tapers off in temperate climates less conducive to their growth, and is a rarity toward the poles. By far the largest consumers are pregnant women, according to Cornell geophagy expert Sera L. Young, author of the book *Craving Earth* and leader of the team that conducted the survey. This makes sense, in her view, for pathogens and dietary toxins pose the gravest threat when cells in the body are rapidly dividing, as happens during fetal development. During the first trimester of pregnancy, moreover, it's not just the fetus that's at high risk of being harmed by these agents but also the mother, whose immune system is suppressed to prevent her from rejecting the fetus growing inside her. Not surprisingly, women in India and Africa often report that they first realize they're pregnant when they develop an insatiable lust for dirt, and many swear by it as a treatment for morning sickness. School-age children are the next-largest population of clay consumers, probably because rapid cell division during growth spurts makes them more vulnerable to pathogens and toxins.

If this accolade to dirt has you hankering to try some, you can purchase it online, as I did, from a favorite supplier of geophagists, Grandma's Georgia White Dirt (www.whitedirt.com); it will be discreetly delivered to your home in an unmarked package. No doubt as a legal hedge, the company markets it as a novelty item not for human consumption. After opening the parcel, I found what looked like small rocks dusted with a fine white powder. With a little trepidation, I bit into one. It was like chewing on chalk, only oilier in texture. Georgia white dirt is definitely not my idea of a mouthwatering treat. After one or two swallows, I spat it out and rinsed my mouth with water — not once but several times, for I could not seem to get rid of its filmy residue on my tongue. Even after I used a toothbrush to scrub it clean, the oily residue lingered, explaining how it coats the GI tract so effectively, creating a long-lasting barrier to parasites. Perhaps I would have found Georgia white dirt more palatable if I'd had a bellyache or if my mother had consumed it while pregnant with me. I suspect a love of clay, like a fondness for spices, may be cultivated in the womb.

Pregnant women who share my disdain for dirt may be comforted to hear that morning sickness may itself be a defense against parasites, so if unavoidable, it need not be a cause for concern. During the first trimester, more than 60 percent of women suffer from it, and the number-one food that causes nausea and revulsion is the dietary item most likely to be contaminated with pathogens — namely, meat, including fish and poultry. Pregnant women's famous aversion to hot cuisines is also thought to be adaptive, for the very property that makes spices inhospitable to pathogenic microbes — their toxicity — could harm the fetus. Morning sickness, in other words, may be nature's way of discouraging an expectant mother from consuming foods that might endanger both herself and her developing baby. In keeping with that theory, such aversions are associated with a lower incidence of stillbirths.

Throughout this chapter, I've demonstrated that humans are not unusual in having a native talent for healing and staying well. However, please note, we are most definitely unique in having the most ad-vanced behavioral defense system of any creature on earth — indeed, our capacity for detecting and dodging parasites without any help from modern medicine is truly stunning. The next chapter will focus on this ability — or feeling, to be more exact. It protects us from both tiny parasites and, more surprisingly, human-size ones — parasites on society, if you will.

A "zombie ant" infected with a parasitic fungus called *Ophiocordyceps*. While on the ground, an ant picks up fungal spores, which send out shoots that invade its body. The fungus then instructs the insect to mount vegetation and lock onto the underside of a twig or leaf. Once the ant is frozen in place, the fungus sprouts from its head and forms fruiting structures that burst, raining spores down on ants below, and the cycle repeats.

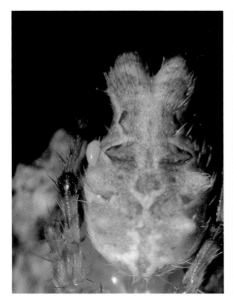

Photos courtesy of William Eberhard

A parasitic wasp has attached its egg to a spider's abdomen (top). As the egg grows into a larva, it injects chemicals into the spider that cause it to replace its normal web (top right) with a more free-form design (bottom right) that will serve as the wasp's nursery. At the center of the customized web, the spider even weaves a decoration that will camouflage the wasp's larva from its enemies. The larva then kills the spider and hooks itself to the web, where it will molt and emerge as an adult wasp.

Courtesy of Marta I. Sanchez

The red "clouds" are swarms of tiny brine shrimp infected with tapeworms, which prod them to seek out other infected shrimp, creating a rich seafood broth that can be easily scooped up by flamingoes — the parasite's next host.

A female jewel wasp performs neurosurgery on a roach. The procedure robs the roach of free will, so when the wasp pulls on one of its antennae, off it trots to her burrow, where it will nourish her offspring.

The yellow sac on the underside of this crab holds the young of a parasitic barnacle. It has also infiltrated the inside of the crab and controls it like an amphibious robot.

Sacculina carcini on the shore crab *Carcinus maenas*. Watercolor by Beth Beyerholm, courtesy of Jens T. Høeg and Jorgen Lützen, University of Copenhagen

Impalas use their teeth like a comb to scrape ticks off each other's necks, a spot the animal cannot reach when grooming itself. This behavior is "programmed": Impalas engage in reciprocal grooming several times an hour even if they don't have ticks (opposite, above right). The red-billed oxpecker (below right) removes ticks deep in their ears — places even a buddy can't reach.

Saliva is rich in antimicrobial compounds that prevent infection —
likely the reason so many animals lick their wounds.

Photos of Ms. Shufa Saleh
Salim (left) and clay (above)
courtesy of Dr. Sera Young:
www.serayoung.org

Clay-rich dirt being consumed by Peruvian macaws (opposite, lower
right), and a Japanese macaque (upper right). A woman (above) gathers
clay-rich rocks in Kenya, where they're sold at open-air markets as a
remedy for gastric distress and morning sickness (inset).

The parasite *T. gondii* can turn a rat's fear of cats into an attraction — part of its scheme for getting back into the feline gut, the only place it can reproduce.

The Czech biologist Jaroslav Flegr believes *T. gondii* can alter humans'
personalities and prompt us to act recklessly.

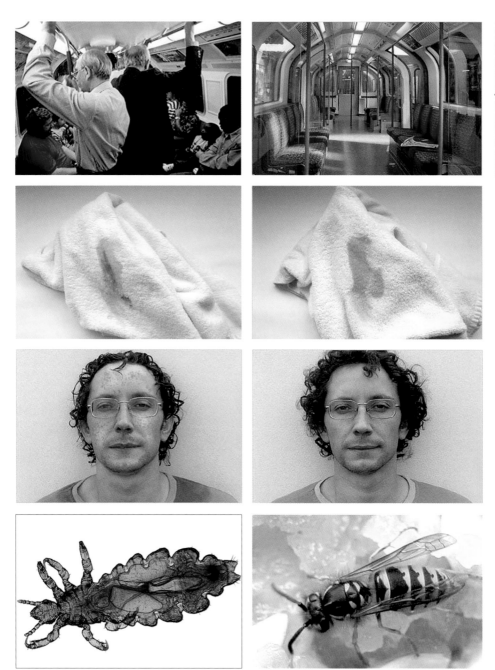

Which subway car will you take? Which cloth will you pick up? Which man will you approach? Which insect repels you the most? Our minds steer us away from sources of contagion by raising our disgust level. In a study conducted by Valerie Curtis in conjunction with the BBC, people around the world overwhelmingly judged each image in the left column more repulsive than its pair on the right — exactly what the researchers predicted based on the salience of disease cues in each picture.

How revolted you are by this image strongly predicts whether you hold liberal or conservative political views. More disgust correlates with a more right-wing stance.

Photo originally appeared in *Predisposed: Liberals, Conservatives, and the Biology of Political Differences* by John R. Hibbing, Kevin B. Smith, and John R. Alford (Routledge 2014). Photo by Anne Nielsen Hibbing.

Rabies has inspired countless vampire legends down through the ages. Not only are vampires capable of morphing into animals that spread the virus, but their hypersexuality is a little-known symptom of rabies in people.

9

The Forgotten Emotion

WHEN VALERIE CURTIS IS NOT in Africa, India, or other parts of the developing world, you'll likely find her at the headquarters of the London School of Hygiene and Tropical Medicine, an elegant old building whose façade is festooned with images of lice, fleas, mosquitoes, and other creatures that have tormented humans down through the ages. In her office, Curtis keeps — tucked up in drawers or occasionally on shelves — many disturbingly lifelike replicas of reviled creatures, vomit, a severed hand, a nasty-looking boil, a rat, and feces. Prominently displayed on her desk is a statue of what looks like a gilded stool.

When my eyes fixed on it, she said, "This is the Golden Poo Award." We were having a video chat, so she hoisted the trophy up to the camera for me to get a better look. At a ceremony that used to be held annually in central London, the Golden Poo was bestowed on heroes of sanitation as a tribute to their efforts to curb diarrheal disease, a leading killer of children in the developing world. Curtis hatched the idea for the award, she said, because she was upset that not enough was being done to bring toilets to poor communities — a failure that she blames on people's reluctance to talk about bodily waste. Delighting her, *Time* magazine wrote about the award and clearly grasped her intention. She paraphrased the article's takeaway: "Which is more shocking, to talk about shit, or the fact that kids are dying because we don't talk about shit?"

In addition to being a crusader for improved hygiene and sanitation in the developing world, Curtis is a self-described "disgustologist" — that is, she's an expert on disgust. The emotion, she and many scientists now believe, evolved to protect us from parasites. Often accompanied by cries of "Yuck!" or "Eww!" and a sense of queasiness and dread, disgust makes us recoil in horror at anything that might sicken us. Poo is taboo, in her view, precisely because it's a pile of germs. Were that not the case, we might not mind its scent and would be happy to talk about it.

Curtis doesn't know if animals experience disgust (it's hard to prove, she points out). She suspects some of them might, that it may, in fact, underpin a number of the behavioral defenses that Benjamin Hart and others have documented in different species. If so, however, animals' limited powers of imagination would greatly curtail its protective value. It's our big brains that make disgust such a powerful germ shield for us.

As we learn more about potential contaminants in our environment, pointed out Curtis, we attach the label *disgusting* to them, and the mere thought of them makes us feel slightly ill and avoid an everbroadening array of hazards. "Disgust is a very sticky emotion," she emphasized.

Still, there's tremendous confusion about this feeling, even as it pertains to humans. For one thing, how we experience it depends on the context. Blood and guts can transmit scores of diseases and consequently are revolting, but if they're encountered on a battlefield they might inspire more terror than revulsion.

To help people sort out the mysteries of disgust and understand it at a visceral level, Curtis brings her plastic turds and ghoulish Halloween-like props into classrooms and other venues. Why is a cockroach disgusting? she asks as she dangles a faux one in front of her audience. Why is this eyeball disgusting? What about this rat?

Surprisingly, most people can't explain their revulsion to these things. A typical answer, according to her, is "I don't know. It's just yucky!"

The question is trickier than it may seem. That's because what disgusts humans is "a really weird mixed bag of filthy, slimy, smelly, sticky, wriggling things," she told me. While some of them — for example, rancid meat, curdled milk, and vomit — are easy to link to illness, in many other cases, the connection is far from obvious.

She and her students have conducted extensive surveys of what people in 165 countries around the globe find disgusting (160,000 participated in one study alone), and many strange items keep popping up in their research.

Acne, for example. It's not contagious, so why is it disgusting? The likely answer is that pimples resemble the pustules associated with diseases like smallpox, measles, and chickenpox.

Rats, cockroaches, snails, and seaweed are widely viewed as disgusting yet none are parasites. They make it onto her list, she believes, because they can transmit viral and bacterial infections, gastrointestinal bugs, parasitic worms, and cholera, respectively.

Earthworms are harmless, but lots of people can't stand to touch them. What makes them repugnant, she thinks, is that they look a lot like parasitic worms in fish and meat that, if swallowed, can burrow into our intestines.

I could go on and on ad nauseam, but I'm sure you'd rather I didn't, so I'll limit myself to just one last example of a weird disgust elicitor: clusters of little holes. The sight of certain patterns of them can make Curtis herself queasy and she is not alone. So many people share this revulsion that it actually has a name: trypophobia. "What sets it off is that the holes have the characteristic arrangement of insect eggs laid in human skin or animal skin," she explained. "It's like a honeycomb pattern." Her voice suddenly sounds wobbly and she starts to wilt right before me. "I don't want to discuss it anymore," she said abruptly. "My skin is crawling, my hair is standing on end." She'd like to study trypophobia, she added, "but I can't bear to. It's so awful."

Are we hardwired to be disgusted by certain things? Clusters of little holes, for example? Or do we learn our disgust from our particular experiences or from the broader culture?

Disgustologists hold strong opinions about this, but firm answers are hard to come by, in part because the scientific community got a very late start in studying the emotion. Over the past century, mountains of tomes were written about anger, depression, fear, and optimism, but disgust, despite its raw, visceral power over us, languished in obscurity. It was called "the forgotten emotion of psychiatry," said Curtis.

Perhaps, I suggested, scientists simply couldn't stomach the subject. Disgust was too disgusting to study. I offered her own reluctance to research trypophobia as a case in point.

"I think you're right," she said. "Also, it was perceived to be a joke area. You can very easily be stigmatized yourself [for studying the topic]. 'Oh, she's the disgust lady.' 'She's that poo lady.'" Even though the London School of Hygiene and Tropical Medicine is devoted to the study of infectious diseases, she had a hard time persuading her colleagues of the area's legitimacy. "Why on earth was I interested in a topic like disgust? they wanted to know. Now they get it. Now they love it. Now they see why I should be researching this area. But when I first started, they thought I was bonkers."

In centuries past, one of the few scientists to take note of disgust was Charles Darwin. In his book *The Expression of the Emotions in Man and Animals,* he described its characteristic face: the corners of the mouth turn down, the tongue thrusts outward, as if expelling something foul. The eyes squint and the nose crinkles, closing off the nasal passages. There's usually an accompanying *Eww,* which is essentially an exhalation: you are pushing contaminated air out of your mouth.

In search of the emotion's evolutionary underpinnings, Darwin wrote to colleagues on almost every continent to inquire how local natives displayed disgust. The reports that he got back led him to conclude that its expression was the same the world over.

Always an astute observer, he was intrigued that merely imagining a revolting idea — for example, swallowing something vile — could actually induce vomiting in some instances. He also made one other very important observation about the emotion: its expression closely mirrors the face a person puts on when displaying contempt for those whose behavior offends. Interestingly in that regard, the responses elicited by Curtis's survey question "What do you find disgusting?" were not limited to the parasitic realm but included corrupt politicians, pedophiles, arrogant Europeans, and wifebeaters, among other commonly loathed groups (Americans may be disturbed to hear that they topped her list of disgust elicitors). As I will detail in later chapters, this tendency to brand offensive people or behavior *repugnant* has major implications for the rise of culture.

Despite Darwin's perspicacity, the only hint that he understood that disgust could guard against infection is a solitary mention of tainted meat being one type of food that commonly evoked revulsion. His uncharacteristic shortsightedness is understandable; his book on emotions was published in 1872, a few decades before groundbreaking experiments by Louis Pasteur and Robert Koch would firmly establish the germ theory of disease.

More than a century would pass before disgust became a focus of scientific study. The psychologist Paul Rozin, venerated as the father of disgust, was one of the first researchers to enter the dormant field. He theorized that the emotion evolved to protect us from food poisoning and bitter toxins but is otherwise largely, if not entirely, determined by culture. He also designed clever and provocative experiments to explore people's sense of contamination. Among my favorites were trials in which he offered subjects items like fudge in the shape of a dog turd or orange juice in which a sterilized cockroach floated. His human guinea pigs were not keen to partake — evidence, he concluded, that our views about contamination are shaped by primitive folk beliefs such as the notion that we might become what we eat or that an object's essence can be imparted to anything that touches it.

Rozin's pioneering investigations of disgust kindled interest in the long-neglected emotion, and his writing on the topic is still immensely influential. But as he himself concedes, his narrow view of disgust's instinctual basis is no longer ascendant. If anything, the pendulum has swung in the opposite direction, as evolutionary psychologists and neuroscientists have flocked to the field. In the opinion of this set, we are innately repulsed by much more than foul-tasting food. Just how many of these core disgust elicitors there are or how best to classify them is still a matter of debate, but when humans encounter them, say scientists in this camp, withdrawal from them is rapid and automatic, rather than reasoned. "There is nothing rational about 'Yuck!'" said Curtis, whose research has been pivotal in bringing about this revision in thinking. Furthermore, culture alone cannot easily explain why so many things that revolt people everywhere — including those in poor remote regions who know nothing of germs — share a link to contagion.

While there are clearly sharp differences of opinion about disgust, some of the dissent between opposing camps may be more superficial than substantial. The folk beliefs that Rozin thinks underlie people's sense of contamination, for example, could well be built on instinct. After all, we can reflect on our behavior and impulses in a way other animals cannot. Indeed, Harvard psychologist Steven Pinker calls disgust "intuitive microbiology," noting that "germs are transmissible by contact," so "it is not a surprise that something that touches a yucky substance is itself forever yucky."

However findings in the field are framed, no one — certainly not Curtis — is suggesting that we come into the world with fixed levels of disgust that thereafter remain inflexible to outside input. "Obviously life experience makes a difference," she said. "Obviously culture makes a difference. Obviously your own personality makes a difference. And so, given the same experience, two people might respond so differently."

Curtis's views on disgust are eloquently set forth in her book *Don't Look, Don't Touch, Don't Eat*, which weaves cultural and biological perspectives into a coherent and compelling picture of the emotion. In essence, she sees disgust as similar to the human sex drive. Individuals vary in the strength of their libidos and what sexually excites them. This drive and how it is manifested changes over one's lifespan and can be altered by unique experiences and societal values. So it is with disgust.

Like our sex drive, the urge to retreat from potential contaminants is not evident at birth but emerges later in development — in the case of the disgust response, shortly after the toddler years, possibly because that's when we begin navigating the world independent of our parents. The trait is then shaped by our encounters with revolting things. If you found a dead rat in your toilet or stumbled upon a badly decomposed body in a forest, you might become more repulsed by rodents or corpses than the average person. Or if, after consuming *moules marinière,* you fell violently ill, the very thought of mussels might make you nauseated for many years to come. In her view, this prolonged aversion to anything we eat that's closely followed by gastric distress reflects nature's unforgiving stance toward those who, upon recovery, return for second servings.

By the time we reach adulthood, there will also be a cultural overlay to what disgusts us — especially in the dietary sphere — but this, too, she argues, is best understood within a Darwinian framework. While dog stew, deep-fried crickets, and whale blubber are relished in some parts of the world and vomit-inducing in others, there's a pattern — an evolved pattern, she believes — to the stupefying variety of regional tastes. You are predisposed to prefer your own culture's culinary customs because following them, at least in the distant past, favored survival in the local habitat, which had its own unique variety of edible plants and animals and traditions for safely preparing them (the use of spices, though not specifically mentioned by Curtis, would be an obvious example of such a practice). If an ingredient is unfamiliar — especially if it's any kind of meat, the food most likely to spoil —

you sniff it with great apprehension before taking the tiniest bite, and if anything about its smell, flavor, or texture raises the least concern, you put down your fork. When in doubt, Curtis believes, the mind's default position is "Better stick with Mama's cooking!"

Standards of cleanliness and sexual mores also strongly influence the spread of disease and, like diet, vary tremendously across the planet. But beneath that diversity, Curtis again sees an evolutionary trend: In every region of the world, people are repelled by poor hygiene and unrestrained sexual behavior at the highest risk of spreading infection. Experiments that tap how disgust affects sexual practices provide additional support for her views. After being shown pictures of people coughing, cartoonish germs sprouting from a sponge, and similar infection-evoking images, women — the sex most vulnerable to STDs — endorsed more conservative sexual values than those who were reminded of other types of threats. And both men and women expressed a stronger intent to use condoms during sex if, as they filled out a survey, they were clandestinely exposed to a rank odor.

Personality adds another wrinkle to the story of disgust. We all know real-life versions of Pigpen from the *Peanuts* cartoons, people who are not much bothered by dirt, two-day-old pizza moldering on the kitchen counter, or the stench of their own bodies. At the opposite pole are fastidious types who shower three times a day, carry around disinfectant wipes, and refuse to use public lavatories. Curtis believes all of us inherit a mixture of genes from our parents that determines where we fall on this spectrum and perhaps even how strongly we're repelled by specific categories of common disgust elicitors — say, blood and guts, sick people, and insect vectors of disease.

Too little or too much disgust can become a liability, so outliers on either end of that continuum will likely be culled from the gene pool. Grown-up versions of Pigpen, for example, will be vulnerable to infection, and their bad breath, stinky armpits, and lice-infested hair will likely scare away potential mates. The easily disgusted, by contrast, may turn up their noses at viable sources of protein — not a good idea when food is scarce — and be at risk for obsessive-compulsive disorder, which in half of cases takes the form of intrusive thoughts about germs and excessive cleaning. Such people may also be unable to countenance the physical intimacy of sex, with its messy exchange of fluids, which could further jeopardize their chances of passing down their genes. Indeed, one-third of people with untreated OCD are either virgins or have not been sexually active for many years. Some cases of agoraphobia (a fear of crowded spaces) and extreme shyness or discomfort in social situations may similarly be "disorders of disgust," Curtis suspects. Still other conditions that may warrant that label are blood-injection-injury phobia (a pronounced fear of having blood drawn with a syringe or having drugs injected into the body) and trichotillomania (a compulsion to pluck out one's hair, which evidence suggests may be triggered by a fear of ectoparasites and an overwhelming urge to remove them).

If Curtis is right, your sensitivity to disgust may even help to determine one of the major dimensions of personality measured by the widely used Big Five personality test. Called neuroticism, the trait is strongly linked to anxiety and depression. Why a characteristic seemingly so detrimental should escape the pruning hand of natural selection has long been a source of puzzlement. Disgust, Curtis suspects, may provide a partial answer to this conundrum. People who score high on neuroticism are averse to taking risks and constantly scan the horizon for signs of trouble — a mind-set that might make them quick to retreat from anyone who shows the least hint of illness and wary of handling grimy objects or consuming food past its prime. Of course, you'd expect neurotic personalities to be leery of many other types of threats too, from accidental injuries and menacing predators to humans brandishing weapons. Few of these dangers, however, would have posed as huge a hazard in the past as the enemy that attacks from within. So if disease avoidance truly counts among the perks of neuroticism, then the trait's cost-benefit ratio suddenly becomes much more favorable. Translation: one possible reason that mood disorders are so common today is that our worrywart ancestors were good at dodging parasites — and they passed their anxious dispositions down to us.

Shining a bright light on psychiatry's forgotten emotion, Curtis predicts, will not only enhance understanding of the forces that shape our personalities but also prove to be a boon for the mentally troubled. People whose agoraphobia or sexual dysfunction stems from a hypersensitivity to disgust, for example, might benefit from very different therapeutic interventions than those whose symptoms — though seemingly identical — have other origins.

Interestingly, women are more sensitive to disgust than men, probably because — in Curtis's words — our female ancestors "had a double burden, to protect themselves and their dependent children from infection." Perhaps related to that observation, women are also more likely to suffer from OCD, social anxieties, phobias, and mood disorders.

Our disgustability and associated vulnerabilities may be affected not only by gender, genetic makeup, and life experiences, but also by the strength of other drives. Hunger is a famous flavor enhancer, so if you live in a place where food is in short supply, say, near the Arctic Circle, even dried strips of putrefying shark meat can taste good — assuming they're prepared properly (Icelanders call this treat *hákarl*). If circumstances are dire enough, you might even eat *hákarl* that's *improperly* prepared, as natural selection has prioritized drives based on the most pressing threat to survival.

Just as hunger can sometimes overpower disgust, so, too, can lust — an adaptation that may have been essential to help our ancestors overcome any reservations about mixing body fluids during reproduction. In support of that view, Curtis points to an experiment conducted at the University of California at Berkeley in which male students were asked to predict how much they'd enjoy having sex in several scenarios. Next, they were instructed to masturbate nearly to the point of climax and again provide responses to the same questionnaire. Sexual acts that they'd previously found off-putting suddenly became much more appealing in their aroused state. The number interested in having intercourse with an obese woman, anal sex, and bestiality, for example, shot up 11 percent, 67 percent, and 167 percent, respectively.

Similar findings have been reported in women. In a Dutch study carried out at the University of Groningen, for example, female subjects were shown a sexually arousing film while the control group viewed clips of high-adrenaline sports like skydiving. Those who saw the erotic film were less repulsed than controls when asked to simulate handling a used condom, rubbing lubricant onto a vibrator, or cleaning a sex toy.

Disgustology is now expanding in numerous directions, offering still more insights into human nature — including a strange quirk that I've noticed in myself and that you may identify with too. In my capacity as a science writer, I have looked over the shoulders of trauma surgeons as they unraveled yards of a patient's gut in search of a bullet wound, all the while not feeling the least bit queasy. Yet seeing a teenager on TV get her tongue pierced so horrified me that I had to flee the room.

The huge disparity in the intensity of my disgust in these two situations is not unusual. A team of researchers led by anthropologist Daniel Fessler at UCLA has shown that seeing violations to appendages repulses people much more than witnessing trauma to organs usually buried deep within the body. For example, subjects rate transplantation of the tongue, anus, or genitalia as far more repulsive than transplantation of kidneys, arteries, or hips. There's an evolutionary reason for this, said Fessler. The parts of the body that interface with the outside world are the most susceptible to damage and infection, so disgust guards us by making us most squeamish about these types of injuries.

How repulsive you find a situation also depends on your familiarity with a potential source of contaminants. As the Brown University psychologist Rachel Herz wryly comments, people admire their own bowels. Your own dirt doesn't bother you as much as other people's dirt. You might not mind sharing your toothbrush with your spouse, but you wouldn't dream of using the toothbrush of a stranger.

The reason for the double standard is that you're immune to your own germs, and chances are you've already been exposed to the germs of an intimate other, so they're not likely to hurt you either. Consequently, the filth and bodily emanations of those furthest removed from your social circle evoke the most revulsion.

Disgust can bias our perceptions in other ways as well. "Imagine brushing your teeth with charcoal-colored toothpaste," Harvard psychologist Gary D. Sherman instructed his audience at a scientific conference. I was among the attendees, and the idea immediately made my stomach turn. In little more than a second, Sherman had effectively made his point — namely, we associate dark colors with dirt and contamination. White, by contrast, signifies purity and cleanliness, hence its pervasiveness in the towels, bedding, and porcelain sinks of hospitals and hotels. This simple observation made him wonder if disgust tunes the perceptual system, making the easily revolted better at spotting contaminants.

To explore the idea, he and collaborators tested subjects' ability to make subtle grayscale discriminations — for example, identifying faint gray numbers against a white backdrop. This talent, they reasoned, would give people an advantage in detecting a speck of dirt. The higher the disgust sensitivity of the subject, they found, the more he or she excelled at the task.

The results were clear: People strongly prone to disgust really did see that speck of dirt around the sink drain better than everyone else. As to why, Sherman was less sure. One possibility is that they're more motivated to hone skills that will help them avoid contaminants. Perceptual tuning that may be comparable has been seen in other sensory modalities, he notes. For example, people who could not initially distinguish between two odors learned to do so when the presentation of one scent was paired with an electric shock. Or causality may run in the opposite direction — that is, the ability to see impurities invisible to others may make people more easily disgusted in the first place. Either way, their world clearly looks very different from that of folks who feel the emotion less intensely.

High levels of disgust may also lead people to invest more time and brainpower to tracking contaminants — real or imagined — when they come in contact with a new surface. Yale neuropsychologist David Tolin swished a pencil around a clean toilet bowl. He then touched the pencil to another pencil, that pencil to still another pencil, and so on. By the fifth pencil, people not troubled by OCD ceased to be concerned about contaminants. Subjects with the disorder, however, viewed even the twelfth pencil as still posing a germ threat.

Whether or not you have OCD, you are probably tracking the spread of contaminants in your environment far more closely than you realize and in situations where the risk of contagion is much less obvious than in the above scenario. Proof of this comes from experiments that explore people's purchasing habits; they show that no one likes to buy anything that has touched another body. For example, clothes on a rack in a dressing room are less likely to be purchased than the same items found on the main sales floor. Consumers even avoid buying items placed near a product with icky connotations. Grocery shoppers, for instance, have been shown to be repelled by foods — including goodies like cookies — if those items come within an inch of touching garbage bags, diapers, or other products associated with filth or bodily waste. Our built-in GPS for tracking germs is evidently very precise at gauging distances — though not always as accurate in flagging true dangers!

Perhaps the most fascinating aspect of disgust is its higher symbolic meaning. This is what most excites Paul Rozin, the founding father of the field, and he and his disciple Jonathan Haidt, now a psychologist at New York University, have arguably made their greatest contribution in this area. Succinctly encapsulating their views, Rozin told a publication of the University of Pennsylvania, where he's a member of the psychology faculty, "Disgust . . . develops from a system to protect the body from harm to a system to protect the soul from harm."

When I spoke to him, he elaborated on this theme. The emotion's most important function, he said, is to shield us from an unsettling truth. We alone among animals know that we will one day die. The thought of decomposing flesh, of worms wriggling through our corpses, is so repulsive that we banish the idea from our heads. Disgust helps us cope with an existential crisis that would otherwise paralyze us. At the deepest level, said Rozin, disgust is about "the denial of death."

This ideational component to disgust — the part concerned with the purity of the soul and our mortality — has extended its reach into numerous spheres of our lives, affecting everything from whom we accept into our social circles to our laws and ethics. A surprising amount of good has sprung from our species' deep-seated fear of contagion — civilization, Curtis thinks, might be one of its byproducts — but there's no denying that it has brought out the worst in us too.

The bad news first . . .

10

Parasites and Prejudice

PARASITES HELD NO INTEREST for Mark Schaller at the outset of his career. Since his graduate-school days in the 1980s, the University of British Columbia psychologist has wanted to understand the root causes of prejudice. In one study that he conducted in the early 2000s, he showed that simply turning off the lights in a room made people more prejudiced against other races. Subjects' heightened sense of vulnerability in the dark seemed to elicit these negative biases — "a relatively obvious idea," he admitted. Then an odd thought occurred to him: "People are potentially vulnerable to infection. Wouldn't it be cool and novel if we found out that prejudices are jacked up when people are more vulnerable to disease?" Or maybe, he thought, he could just disgust subjects with "a cute manipulation" (about which more in a moment) and see if their attitudes toward outgroups — those perceived to be racially or ethnically different from themselves — shifted in a negative direction.

Schaller, who approaches science with a playful spirit ("I love wild ideas," he told me), had no qualms about entering the forbidding realm of the yucky because he's not easily disgusted. When pressed to elaborate, he shared a story about a meal that he'd prepared at his home for

Paul Rozin and his wife. In the middle of it, Schaller spotted "a fairly sizable beetle" that had landed on his own plate, evidently having ridden in on some raspberries that he'd picked from his garden earlier that day. "I pointed it out, since this was a classic food-and-disgust kind of thing that Paul would appreciate, and his wife said, 'The question is: Are you going to eat it?'

"Of course I said yes. How could I not in those circumstances?"

He did not regret it. "I was not disgusted at all," he insisted, adding, "I've put a slug in my mouth before. I don't need any provocation to do it."

I mention Schaller's robust constitution because I suspect it may explain his hubris in attempting the "cute manipulation" earlier alluded to. His plan was to test whether subjects developed more negative views of outgroups immediately after they ate a durian. For those unfamiliar with this exotic fruit from Southeast Asia, a durian looks like a prickly football and has edible flesh that is famously stinky. Indeed, its stench is often likened to that of rotting onions or sweaty gym socks — mountains of them (*overpowering* is a common adjective used to describe its odor).

"I went to a little Vietnamese market to buy one," said Schaller, "and they almost refused to sell it to me. They said, 'Do you realize what you're getting here?'" Not dissuaded, he proceeded with the transaction, but even he was no match for the durian. On the drive home, he admitted, "I had to put it in the trunk."

Alas, he weathered the assault on his senses for no apparent scientific gain. The fruit that made Schaller gag turned out to be an unreliable elicitor of disgust. Vancouver, where the study was conducted, has a huge Asian population, and a number of the subjects in the trial were familiar with the durian and loved it. According to him, their reaction was "Yeah, it's stinky — but it's delicious!" He was forced to scrap the trial, but before doing so, he noticed that the data collected from non-Asians suggested that his theory — far-fetched as it sounded — might have merit.

Schaller tried another tactic, opting to revolt subjects with a slide show of snotty noses, faces covered in measles spots, and other disease-related stimuli that previous research had demonstrated evoked near-universal disgust. The control group saw pictures depicting threats unrelated to infection — for example, electrocution or being run over by a car. All the subjects were then asked to fill out a questionnaire that assessed their support for allocating government funds to help immigrants from Taiwan and Poland (groups whom they ranked as very familiar, as Vancouver is also home to many transplants from Eastern Europe) versus immigrants from Mongolia and Peru (whom they rated as unfamiliar). In comparison to the controls, the subjects who saw the germ-evoking photos showed a sharply elevated preference for familiar immigrant groups over lesser-known ones.

Drawing on more than a decade of research by himself and others since the study was published, Schaller offered this interpretation of the findings: Over human history, exotic people have brought with them exotic germs, which tend to be especially virulent to local populations, so foreignness seems to trigger prejudice when we feel at greater risk of getting sick. Also, it may be that lurking in the backs of our minds are concerns that the foreigner does not have as high standards of hygiene or that he doesn't follow culinary practices that reduce the risk of food-borne illness. Prejudice, Schaller points out, is all about shunning others based on superficial impressions, so the sentiment, ugly as it is, is ideally suited for the purpose of shielding us from disease.

Related trials suggested that the mind's sense of "foreign" is blurry. Schaller, in collaboration with other researchers, discovered that any reminder of our susceptibility to infection makes us more prejudiced against the disabled, the disfigured, the deformed, and even the obese and elderly — in short, a vast swath of the population who pose no health threat to anyone.

"Infectious disease causes a wide variety of symptoms so we're probably picking up on the fact that the person is not looking normal," he said. By *normal*, he means a caveman's notion of what a healthy person should look like. Until very recently, "the prototype human being" — as he puts it — was rarely overweight or much older than forty, so people who are obese or show signs of old age, like bags under the eyes, liver spots, and curled yellow nails, are categorized as weird. Like a smoke detector, your germ-detection system is designed to sound at the least hint of danger. A false alarm could mean a lost social opportunity, but if someone displays contagious symptoms that you mistakenly think are innocuous, it could cost you your life. "Better safe than sorry" seems to be nature's motto.

The germ-detection system is not only crudely calibrated but also designed to operate largely outside of conscious awareness, being much more swayed by feelings than facts. To underscore that point, Schaller described an experiment that began with his team showing subjects photos of two men. The first had a port-wine birthmark on his face but was characterized as strong and healthy. The second man looked outwardly robust but accompanying information told the subjects that he had a very contagious drug-resistant strain of tuberculosis. The participants were then given a computer-based reaction-time test to tease out which of the men they subconsciously associated with infection. In spite of the information shared with the subjects, the test revealed that they perceived the man with the harmless birthmark to pose the greater disease threat.

According to Schaller, subjects also stare longer at disfigured faces than they do at the same faces photographically altered to correct the abnormality. In cognitive science, as a rule, the more attentive you are, the better your recollection. For example, people peer longer at angry faces and remember them better afterward. But in the case of disfigured faces, the reverse effect was noted: participants in the trial had much poorer recall of individuals displaying anomalies, often confusing them with one another. Or as Joshua Ackerman, a scientist who was involved in the trial, put it, the participants in the study "were looking without really seeing."

"They all look the same to me" is a common refrain heard when people are called upon to tell apart individuals who belong to unfamiliar races, and the same dehumanizing trend seems to apply here.

In the case of an angry face, we carefully encode the features so that we'll recognize a potentially hostile person in another situation. But the unique traits of a disfigured person — save for the disfigurement itself — are of no use in tracking a potential source of contagion, with the result that we focus on the salient threat to the exclusion of the individual's other characteristics.

To Schaller, it's "mind-boggling" that scientists have only recently come to appreciate that parasites in our surroundings might inflame prejudice, given that they've known about other behavioral defenses against disease — especially in animals — for decades. Viewed from a different angle, however, the oversight didn't surprise him. "A lot of what people study is based on their own personal experience, and most work in the psychological sciences is done in Canada, the U.S., and Europe in places like this," he said, casting his eyes about. We were seated in a sparkling new building on the UBC campus with austere modern lines and sleek, minimalist décor — about as sterile a setting as one could imagine. "We don't really worry much about infectious disease. We forget that in most of the world and throughout most of our history, infectious organisms have posed this extraordinary health threat and have almost certainly played a huge role in human evolution, including the evolution of our brain and nervous system."

Schaller coined the term the *behavioral immune system* to describe thoughts and feelings that automatically spring to mind when we perceive ourselves to be at risk of infection, propelling us to act in ways that will limit our exposure. As his own research was advancing, so, too, was the work of scientists like Rozin, Haidt, and Valerie Curtis; consequently, he told me, the label was chosen to encompass not just germ-induced prejudices but also a broad repertoire of other disgust-based protective responses against infection as well as behaviors in animals that serve the same function.

While he clearly thinks insights from this domain have much to teach us about interpersonal relations, he's careful not to oversell his findings. A subconscious fear of contagion, he underscores, is hardly the sole cause of prejudice. We may negatively stereotype different races or ethnicities out of anger that they may threaten our livelihoods or out of fear that they may want to harm us. We may shun the disfigured and deformed because they are reminders of our own vulnerability to injury and misfortune. Or prejudice may simply be born of ignorance — the denigration of the obese as lazy and slovenly, for example, may stem from someone having little contact with overweight people in a professional setting. Even if we could banish the world of infectious disease, said Schaller, it wouldn't eradicate prejudice.

He offered an additional caveat: "A lot of the research we've done has focused just on our initial automatic response to people who activate our behavioral immune system, but that doesn't mean that's all that is going on in our heads. For example, my initial response to someone who is weird-looking might be revulsion, but that may be immediately superseded by a more deep sympathetic response that takes into account the predicament the person is in and can elicit sensitivity and understanding. These additional, more thoughtful responses may not be the first things that cross our psychological radar but they may ultimately have a much bigger effect on how we respond in real life in that situation."

Nonetheless, studies by Schaller and other researchers indicate that people who chronically worry about disease are especially prone to antipathy toward those whose appearances diverge from the "normal" template, and these people have a harder time moving beyond that reaction. This can have real, long-lasting effects on their attitudes and experiences. Compared to people not plagued by such health concerns, they are less likely to have friends who are disabled; by their own accounts, they are less inclined to travel abroad or engage in other activities that might bring them into contact with foreigners or exotic cuisines, they more frequently display negative feelings toward the elderly on tests of implicit attitudes, and they report harboring greater hostility toward the obese. Indeed, the more they fret about getting sick, the greater their expressed disdain for the obese, possibly explaining why fat people are so frequently branded with pejorative adjectives strongly linked to infection, such as *dirty, smelly,* and *disgusting.*

These antipathies affect how germaphobes interact with everyone, not just strangers. Parents prone to such fears report having more negative attitudes toward their fat children — sentiments that don't carry over to their normal-weight offspring.

The recently ill display similar biases, possibly, Schaller theorizes, because their immune systems may still be run-down, so their minds compensate by ratcheting up behavioral defenses. In support of that contention, he points to a provocative study by evolutionary biologist Daniel Fessler and colleagues, who showed that pregnant women become more xenophobic in the first trimester, when their immune systems are suppressed to prevent rejection of the fetus, but not in later stages of gestation, when that danger has passed. Further research by Fessler in collaboration with Diana Fleischman revealed that the hormone progesterone, which is responsible for reining in the immune system early in a pregnancy, elevates feelings of disgust, which in turn promotes negative attitudes toward foreigners *and* pickier eating habits — the latter response likely an adaptation that discourages pregnant women from consuming foods prone to contamination, as we saw in chapter 8. In other words, it appears that by evoking disgust, a single hormone initiates two behavioral defenses at exactly the time in pregnancy when the danger posed by infection is greatest.

Such hormone-induced shifts in feelings are not limited to gestation. During the luteal phase of a woman's menstrual cycle (the days that follow the release of an egg from her ovaries), progesterone rises to allow an egg, should it become fertilized, to implant in the womb without being attacked by immune cells. By measuring salivary levels of the hormone in regularly cycling women, Fessler and Fleischman discovered that the luteal phase is accompanied by heightened feelings of disgust, xenophobia, and concern about germs. For example, women at that stage in their cycles reported more frequent handwashing and use of paper seat liners for toilets in public restrooms.

"Understanding the sources of some of these attitudinal changes is potentially important," said Fessler. "In teaching my undergraduate students about how to understand the mind from an evolutionary perspective, I try to make the point that we're not slaves to our evolved psychology. When a woman walks into a ballot booth to make a decision about a candidate based on his or her immigration policies, for example, this knowledge gives her the power to step back and say, 'Well, wait a minute. Let me make sure my decision reflects my well-thought-out position on this issue, and not impulses I'm experiencing at this moment.'"

THE BEHAVIORAL IMMUNE SYSTEM INFLUENCES more than our attitudes toward foreigners; it also, several studies suggest, affects our gregariousness — and consequently how frequently we come in contact with potential germ carriers. After seeing pictures that call to mind the threat of infectious illness, both male and female subjects reported being reluctant to approach strangers and rated themselves as more introverted — changes not seen in control subjects shown photos of architecture. While this shift in sociality is usually transient, people who habitually fret about falling ill commonly describe themselves as being introverted even in the absence of any immediate infection threat. They also routinely rank themselves as less agreeable and less open to new experiences — traits that promote a more hostile and distrustful stance toward outsiders and their unfamiliar customs.

Still, those who fear contagion — whether chronically or just in high-risk situations — don't shy away from all company. Studies suggest that they're ethnocentric; that is, they predominantly socialize with familiar people, such as family and close associates. One likely reason, researchers believe, may be that this tight circle — the in-group, to use the terminology of social scientists — can be relied on for care and support should the germaphobe fall ill.

Intriguingly, several studies indicate that lowering concerns about the risk of infection — for example, by vaccinating subjects — can effectively shut down the behavioral immune system, suppressing prejudicial feelings that would otherwise arise when the threat of contagious disease is made salient. In one notable experiment, being allowed to use an antibacterial wipe to disinfect hands when reminded that the flu was going around decreased negative attitudes toward a variety of perceived outgroups, including illegal immigrants, Muslims, the obese, and the disabled — and the reduction was most dramatic in the germaphobic. Ackerman, one of the experimenters, aptly refers to this phenomenon as "washing away prejudice."

Political scientists are now streaming into the field, testing whether central findings hold up across different cultures and in populations much larger than typically studied by psychologists. One of the biggest and best-controlled of these trials, conducted by Michael Bang Petersen and Lene Aarøe, included nationally representative samples of two thousand Danes and thirteen hundred Americans whose vulnerability to infection was assessed by multiple measures. The Danish subjects took an online test that ranked their sensitivity to disgust, and information was collected from them about the last time they'd been ill and how often they worried about infection. Next they completed a test designed to reveal xenophobic tendencies. The scientists' findings dovetailed neatly with Schaller's results from laboratory studies, showing, among other things, that opposition to immigration increased in direct proportion to disgust sensitivity.

In their analysis of the American sample, the political scientists took their research one step further. They collected data from the CDC on the incidence of infection on a state-by-state basis and cross-correlated those figures with the number of online searches conducted in each state using key words related to contagious disease (Google Trends gathers this information, which allows epidemiologists to predict the onset of flu epidemics and other disease outbreaks). After controlling for every variable the researchers could think of — race, age, gender, education, socioeconomic factors, the state's unemployment rate, the size of its immigrant population, whether it was red or blue — they found greatest opposition to immigration where contagious disease was most prevalent and, predictably, where worry about infection was highest.

The last segment of their study proved the most enlightening. The American subjects were given information about a male immigrant from either a familiar country or a more foreign culture; one group was told he was very motivated to learn English and committed to American values of democracy; another group was told he was not very motivated to learn English and skeptical of American ideals; and the third group received no information about his desire to fit in. Participants who worried about germs strongly favored rolling out the welcome mat for the immigrant of familiar origin over the one from the lesser-known country, and this held true regardless of whether the newcomer was likely to contribute to society or embrace its values.

"This suggests that the behavioral immune system is relatively robust and insensitive to the kind of information that we know otherwise promotes peaceful coexistence and tolerance," said Aarøe.

Her collaborator Petersen added, "If what I'm fearing is not so much what you will do to me but what your pathogens will do to me, then your intentions become irrelevant. That says something crucial about why ethnic integration is so difficult to achieve."

The scientists' findings fly in the face of current wisdom, which emphasizes that striving to fit in is the key to assimilation. Their results also highlight how poorly our germ-tracking radar is adapted to protecting us from disease in a modern world in which people of diverse ethnicities often live side by side. What's more, the elderly and obese make up a large proportion of contemporary societies, presumably triggering still more misfiring of the behavioral immune system. "It's hyperactive," said Petersen, and for little protective gain, especially in affluent regions where the risk of infectious disease is far lower than it was in the environment in which our ancestors evolved. "That's one potential implication."

But he and Aarøe don't rule out an alternative, more optimistic interpretation. It could be that an individual's sense of what a "normal" person should look or act like is based on the kind of people he or she interacts with every day, which should help to dial down germ-alert levels and associated prejudices.

Schaller favors that view. "If I grow up in an environment where everybody looks pretty much the same, then someone from China, for example, might trigger my behavioral immune system. But if I grow up in New York City, then a person who comes from China is not going to trigger this response. It's also changeable in other ways too. To the extent that people are aware of real threats to their health, that can reduce these prejudices."

Unfortunately, however, most of us are blind to the inner workings of our minds and the degree to which our perceptions are biased by evolutionary design. Anti-immigrant propaganda brilliantly capitalizes on the human propensity to view foreigners as sources of contagion. Its ugly rhetoric is familiar: Outsiders are filthy, repulsive, and lice-infested and will contaminate you with no end of vile germs. Against this backdrop, outbreaks of disease are invariably attributed to their polluting presence, especially in earlier centuries, when the true cause of epidemics was unknown and the death toll from infection far higher. Jews, it was rumored, brought the Black Plague to medieval Europe by poisoning the wells of Christians — a baseless claim for which the accused were immolated in numerous pogroms. In more recent times, thankfully, we're not as quick to torch outsiders suspected of sickening us, but we're still inclined to stigmatize them. The world blamed the devastating 1918 flu pandemic on the Spaniards, who in turn blamed the Italians (for the record, it likely originated in Kansas). In the early years of the AIDS epidemic, Haiti was demonized for unleashing the virus on the world — a view not only false but also damaging to the nation's already fragile economy. And in 2014, anti-immigration websites and even some U.S. congressional representatives warned that the stream of Latin American refugees flooding the nation's southern border might infect citizens with Ebola — and this despite the fact that not a single case of the disease had occurred south of Texas.

No matter where you reside, immigration is virtually guaranteed to be a hot topic come election time for reasons that are by now probably obvious — it's a perennial vote winner. Of course, there are arguably legitimate reasons for opposing immigration, but they gain added weight when people are already predisposed to perceive the outsider as a germ threat, especially when politicians stoke those very fears.

Xenophobic propaganda can take another sinister form. Its progenitor is that favorite taunt of the playground bully: "You've got cooties." Grown-up bullies are notorious for fomenting hatred by branding the target of their aggression — usually a vulnerable minority — a parasite or other vehicle for transmitting infection. This tradition has deep roots. The ancient Romans vilified outsiders as detritus and scum. Jews — history's favorite scapegoats — were depicted by the Nazis as leeches on society, setting the stage for the Holocaust. Meanwhile, in the United States, law-abiding Japanese American civilians were called "yellow vermin" — a slur that became a rallying cry for imprisoning them in internment camps. In 1994, Rwanda erupted in a genocidal bloodbath when Hutu extremists incited their followers to "exterminate the Tutsi cockroaches."

White supremacists have an extra advantage in leveraging our fear of contagion for the purpose of fueling hate. Since dirt, feces, and many insects that transmit disease are typically brown or black, many of us associate dark shades with impurity and contamination. Such habits of thought may well have helped to justify banning blacks in the Jim Crow South from sharing lunch counters and drinking fountains with whites — and, an even greater anathema to white southerners, sharing swimming pools. Indeed, some of the most bitterly fought battles over integration took place at pools, and anger over the issue followed blacks on their migration north. In Pittsburgh in the 1930s, African Americans were dragged out of a public pool and ordered to stay away unless they could proffer health certificates showing that they were free of disease. A generation later, Latinos on the West Coast faced similar persecution: In some Los Angeles communities in the 1950s, Hispanic Americans were allowed to swim only on Mondays, "Mexican Day," after which the pool was drained and refilled for use by whites alone. Gerald L. Clore and Gary D. Sherman, the psychologists who studied the mind's tendency to link dark colors with impurity, suspect fear of contagion might even lie at the root of anti-miscegenation laws in the South; the one-drop-of-blood rule, which remained in effect until the mid-twentieth century, held that a mere trace of African heritage could taint the "pure" white race.

The implication of this research borders on the absurd. Each of us, it suggests, might be able to enhance our acceptance and integration into society through special — some might even say obsessive — attention to cleanliness. To put it bluntly, our inner bigot smiles on the immaculately dressed, the well-coiffed (hair ideally tied back from the face), the clean-shaven, the deodorized, the sweet of breath, and the meticulously manicured. To complete this picture of perfect hygiene, accessorize with latex gloves and a surgical mask. Okay — maybe that's going a bit far, but scientists have theorized that wearing medical gear should reduce prejudice.

Or perhaps the onus should be on the easily disgusted to challenge their inner bigots. When you catch yourself pulling back from someone who revolts you, ask yourself: *Does this person threaten my health?* If not, consider leaning in instead — and, if appropriate, offer a handshake or a hug. Admittedly, that's sometimes easier said than done, especially for the readily repulsed. A friend of such a disposition confessed that he was unable to sit next to someone with psoriasis — forget about shaking hands — even though rationally he knew the skin condition posed no risk to him.

Sadly, the sick are the most likely to activate the behavioral immune system, so on top of coping with their illnesses, many are weighed down by the shame of repulsing others. Cancer patients often feel this acutely. After having a double mastectomy in 1975, NBC correspondent Betty Rollin recalled pulling down the shade in her bedroom — not to keep lecherous eyes from ogling her but rather "to keep the mythical peeping Tom from throwing up." The best-selling book she wrote about her experience, *First, You Cry*, nearly didn't get published.

When she pitched the proposal to editors, the response of nearly all of them was "Breast cancer — yuck! Who'd want to read about that?" She wrote the book anyway, bringing the once-taboo topic out into the open and many sufferers out of self-imposed exile.

Some may think that we've since conquered our queasiness about breast cancer and other diseases with ugly reputations, but my own experience tells me otherwise. Only a few years ago, one of my dearest friends developed anal cancer but told colleagues at work she had colon cancer. *Anal cancer sounds so gross!* she wrote me, later admitting that her embarrassment had even delayed her from letting those closest to her know her true diagnosis. More recently, an overseas friend on the verge of succumbing to a respiratory disease declined to turn on his camera during our last conversation over Skype. Afterward, I learned that he wanted to shield me from seeing him gaunt and tethered to an oxygen tank via tubes in his nose.

In the course of her research on disgust and related health issues, Valerie Curtis hears these kinds of stories all the time. The sense of shame is particularly acute for people who are incontinent, she reports. Such people are more likely to be avoided and, in some societies, treated as outcasts. Even those who care for the sick or who have unsavory jobs like cleaning toilets or sewers are frequently stigmatized. Worse, they themselves may develop posttraumatic stress disorder when called upon to perform particularly nasty chores — for example, retrieving the corpses of neglected old people who've been left to rot in their homes. "I think it's terribly important we talk about this," Curtis said, "because it's no good saying, 'Oh, you're being irrational and illogical. Of course nobody thinks you're disgusting because you're sick.' But I'm afraid we do. And we're not going to make much progress unless you acknowledge it and deal with the emotional labor involved in having to overcome disgust."

Encouragingly, research suggests that it may be possible to dial down this powerful emotion through repeated exposure to repulsive stimuli. For example, people who routinely change wound dressings and soiled bedding report reduced sensitivity to disgust, though unrelated disgust elicitors — say, sour milk or slugs — may still make them recoil.

Although research on disgust has grown immensely over the last two decades, important questions remain unanswered — chief among them, does this stomach-churning emotion influence the functioning of our immune cells? In other words, is the psychological immune system communicating with the physical immune system, or do they operate largely independently of each other?

Unfortunately, researching these issues is difficult to do; it's costly and requires expertise beyond the scope of many psychologists. By enlisting the assistance of a team of neuroimmunologists, however, Schaller succeeded in conducting one of the few studies to address the question. As in many of his previous trials, subjects were shown a disease-y slide show, but with one major difference: Immediately before and after the presentation, their blood was drawn and mixed in a test tube with a pathogen surface marker to determine how aggressively their white blood cells countered the challenger. Specifically, the investigators looked to see if arousing subjects' disgust spurred their white blood cells to produce higher amounts of a pathogen-fighting substance called interleukin 6 (IL-6).

It did — and by a whopping 24 percent. By comparison, the control group of subjects, who saw pictures of people brandishing guns pointed directly at them, showed virtually no change in IL-6. Interestingly, said Schaller, although the germ-evoking pictures were much more effective in revving up the immune system, the gun photos were actually judged to be more distressing, demonstrating the specificity of the immune response.

No one has yet attempted to duplicate this trial, so more research is clearly needed to draw reliable conclusions. Still, related studies support the experiment's general findings. In an Australian investigation, for example, researchers collected saliva samples from subjects before and after showing them pictures of vomit, a cockroach on a pizza, and similarly icky fare. Compared to participants who saw pictures of neutral content, the grossed-out subjects produced significantly more tumor necrotizing factor alpha — an immune substance in their saliva involved in fighting infection. A similar British study obtained blood samples from subjects before and after flooding their senses with nausea-inducing images of blood, gore, and dismemberment, a feat accomplished by showing them the gruesome 1974 film *The Texas Chainsaw Massacre*. The participants' leukocytes — white blood cells that fight off infection — climbed steeply in number, a change not seen in the control subjects, who spent the duration of the film reading mundane material.

If, as this research implies, the disgusted mind truly can shift the immune system into high gear, it makes very good sense, in Schaller's opinion. "Our eyeballs are providing useful information to our immune system. If they're telling us there are a lot of sick people or other sources of germs around, that indicates that we ourselves are likely to become exposed or maybe already have been, so ramping up the immune system gives it a head start in fighting off microbial invaders."

He thinks there may be another virtue to this biological setup as well. "The information allows the immune system to calibrate the aggressiveness of your response to the scale of the threat. We don't want the immune system working hard unnecessarily because it consumes a lot of resources that might be used by other parts of the body."

How, at a neurological level, the psychological immune system might "talk" to the physical immune system is still a matter of speculation. But scientists have begun to track where disgust is processed in the brain, and evidence suggests this region, through a little rejiggering of its circuitry, has evolved to serve an important new function — one that arguably defines the very essence of our humanity. As will soon become clear, we may have disgust to thank for transforming our species into the most freakish of creatures: a moral animal.

11

Parasites and Piety

THE YOUNG MAN WAS HAVING SEX with his dog. In fact, he'd lost his virginity to the dog. Their relationship was still very good. The dog didn't seem to mind at all. But the man's conscience was eating at him. Was he acting immorally?

In search of sage counsel, he sent an e-mail to David Pizarro, who teaches a class on moral psychology at Cornell University. "I thought he was just pulling my leg," said Pizarro. He sent the man a link to an article about bestiality and thought that would be the end of it. But the man responded with more questions. "I realized this kid was pretty serious." Although Pizarro is a leader in his field, it was clear that he'd struggled to craft an answer: "What I ended up responding to him was: 'I might not say this is a moral violation. But in our society you're going to have to deal with all manner of people believing that your behavior is odd, because it is odd. It's not something anybody likes to hear about.' And I said, 'Would you want your daughter to date someone who has been having sex with their dog? And the answer is no. And this is critical: You don't have animals writing essays about how they've been mistreated because of their love of human beings. I would get help for this.'"

In essence, Pizarro was saying the man's behavior was weird, concerning, and distressing, but he wasn't willing to condemn it. If that doesn't sit well with you, no doubt you're sickened by the very image of someone having sex with a dog. But was the man truly acting immorally? At least by his own account, the dog wasn't being harmed. Who is the victim in this disconcerting tale?

If you're struggling to put your finger on why exactly his behavior seems wrong, psychologists have a term for your confused state of mind. You're *morally dumbfounded.*

A ballooning body of research by Pizarro and others scientists shows that moral judgments are not always the product of careful deliberation. Sometimes we feel an action is wrong even if we can't point to an injured party. We make snap decisions and then — in the words of Jonathan Haidt, another giant in the field of moral studies — "construct post hoc justifications for those feelings." This intuition, converging lines of research reveal, is informed by disgust. Over the course of human evolution, the same emotion that made people gag at a rancid odor or spew spoiled milk somehow became embroiled in our most deeply held convictions, from ethics and religious values to political views.

Disgust's key role in our moral intuitions is echoed in language: Dirty deeds. Slimy behavior. A rotten scoundrel. Conversely, cleanliness is next to godliness. We seek spiritual purity. Corruption can contaminate us, so we shun the evil.

Pizarro has a deep distrust of using disgust as a moral compass. If people rely on it, as they often do, it can lead them astray, he warns. In his class, he offers the denunciation of homosexuality on the grounds that it's repugnant as a prime example of the dangers of disgust-based morality. "I tell my class: As a heterosexual male, it's not as if I won't be disgusted if you show me pictures of certain sexual acts between two males. The task for me is to say: What the hell does this have to do with my ethical beliefs? I tell them, the thought of two very ugly people having sex also revolts me, but that does not lead me to consider legislating against ugly people having sex." The homeless are another group that people frequently speak ill of, probably because they, too, can trigger disgust alarms, making it easier for society to dehumanize them and find them guilty of crimes they didn't commit. "My ethical duty is to make sure that this emotion doesn't influence me in a way it might actually tread on someone's humanity," Pizarro said.

He knows better than most the travails of keeping disgust in its place and not letting it seep into his ethical judgments. He is so squeamish that he has to rely on his students to program all the pictures of repulsive imagery that are used in his studies on moral reasoning. "It took just brute reasoning to shake myself of some of my attitudes," he said. "I consider it an intellectual achievement that I became liberalized about a lot of issues."

The curse of being exceedingly easy to disgust has nonetheless been a benefit in his work. As he acknowledges, it has given him keen insights into how the emotion can guide moral thought.

If you're skeptical that parasites have any bearing on your principles, consider this: Our values actually change when there are infectious agents in our vicinity. In an experiment by British psychologist Simone Schnall, students were asked to ponder morally questionable behavior such as lying on a résumé, not returning a stolen wallet, or, far more fraught, turning to cannibalism to survive a plane crash in a remote location. Subjects seated at desks with food stains and chewed-up pens typically judged these transgressions as more egregious than students seated at spotless desks did. Numerous other studies — using, unbeknownst to the participants, imaginative disgust elicitors like fart spray or a chemical that mimics the scent of vomit — have reported similar findings. Premarital sex, bribery, pornography, unethical journalism, marriage between first cousins — all become more reprehensible when subjects were disgusted.

The repulsed are also more likely to read evil intent into innocuous acts. In a trial conducted by Haidt and graduate student Thalia Wheatley, hypnotic suggestion was employed to cue subjects to feel a wave of disgust when they came across the words *take* and *often*. Then the volunteers read a story about a student council president named Dan who was trying to line up topics for students and professors to discuss. It had no moral content. Yet subjects who got the version that contained the disgust-trigger words were more suspicious of Dan's motives than controls who read a virtually identical story without the hypnotic cues. Explaining their distrust of Dan, people in the experimental group offered rationalizations such as "I don't know, it just seems like he's up to something" and, to Haidt's amusement, "Dan is a popularity-seeking snob."

Harmless sex acts can similarly acquire immoral overtones if germs are foremost on your mind. When subjects in one of Pizarro's studies were shown a sign recommending the use of hand wipes, he reported, they were harsher in their judgment of a girl who masturbated while clutching a teddy bear and a man who had sex in his grandmother's bed while housesitting for her.

There's a clear pattern to these findings, as an investigation by psychologists Mark Schaller and Damian Murray reveals. People who are reminded of the threat of infectious disease are more inclined to espouse conventional values and express greater disdain for anyone who violates societal norms. (Incidentally, concerns about bad drivers, war, and other threats to safety also elicit greater fondness for conformists, but not as dramatically as germ fears do.) Disease cues may even make us more favorably disposed toward religion. In one study, participants exposed to a noxious odor were subsequently more likely to endorse biblical truth than those not subjected to the polluted air.

When we're worried about disease, it appears, we're drawn not just to Mama's cooking but also to her beliefs about the proper way to conduct ourselves — especially in the social arena. We place our faith in time-honored practices probably because the tried-and-true seems like a safer bet when our survival is in jeopardy. Now's not the time to be embracing a new, untested philosophy of life, whispers a voice in the back of your mind — the part that, whether you're conscious of it or not, is constantly assessing the risks and advising how to react in response.

In light of these findings, Pizarro wondered if our political attitudes might shift when we're feeling susceptible to disease. In collaboration with Erik Helzer, he came up with a clever strategy to test the idea. They positioned subjects next to a hand-sanitizer station or where none was in sight and asked about their views on various moral, fiscal, and social issues. Those reminded of the dangers of infection expressed more conservative opinions.

Intriguing as these results are, they should be interpreted with caution. When we're called upon to render moral judgments in real life, we have far more information to go by than in a laboratory setting — among other things, people's demeanor, how they generally comport themselves, mitigating circumstances, and so forth. "There are a lot of influences on moral judgment and disgust is only one of them," emphasized Pizarro. In the more complicated world of everyday life, snap decisions based on visceral disgust are no doubt often later tempered by logic and reason, leading us to modify our initial assessment of an infraction or even conclude it constitutes no breach of ethics at all. What's more, disgust acts on a person's already well-developed system of values. A filthy desk or a whiff of a foul scent does not turn sexual libertines into prudes, atheists into religious zealots, or renegades into conformists. "The shift in attitudes is temporary and modest in size," stressed Pizarro. "I try to emphasize when I give talks on this stuff that if you want to influence people's attitudes, there are probably much more powerful ways to do it."

Those caveats may have bearing on the outcome of a recent study he did that aimed to test whether a particularly robust laboratory finding — the fomenting of anti-gay sentiment in response to disease cues — held up in the real world. Working in collaboration with Yoel Inbar and researchers at the University of Virginia, he and his team conducted an online survey of Americans' attitudes toward homosexuality as concerns about the Ebola outbreak reached a fever pitch during the autumn of 2014.

Implicit views toward the group did indeed shift in a negative direction, the scientists found, but the effect was much smaller than they'd anticipated.

Maybe it was so weak, Pizarro theorized, because participants, even those overwrought about Ebola, may not have been worrying about the disease close enough in time to when their opinions were actually tapped (in the lab, surveys are filled out within a few minutes of exposure to a noxious odor). But he does not discount an alternative possibility — namely, disgust elicitors tend to amplify existing prejudices, and societal attitudes toward gays have radically changed in just the past few years, with this once-reviled group now viewed far more favorably. If that's why the outbreak had such a puny effect on anti-gay sentiment, Pizarro said, "it would be really encouraging."

If you're squeamish by temperament, however, the emotion's impact on your attitudes may be neither minor nor fleeting. Mirroring laboratory findings, research by Pizarro and other groups suggests that the readily revolted are more likely to hold stable political views at the conservative end of the spectrum — and not only regarding immigration, as we saw in the previous chapter. The highly disgustable also tend to be hard on crime; against casual sex, abortion, and gay rights; and authoritarian in orientation. For example, they're more inclined to think children should obey their elders without question and they place greater emphasis on social cohesion and following convention. Though the evidence is not as strong, there are even hints that those prone to disgust are more likely to be fiscally conservative (against taxation and government spending programs).

There's a physiological angle to this story as well. When shown pictures of people eating worms and other revolting imagery, conservatives sweat more profusely than liberals (as measured by galvanic skin response). Their heightened reactivity, however, is not limited to disease-related dangers. Compared to liberals, they also react to loud noises with a more pronounced startle response. These twin observations may have direct bearing on a well-documented finding in political science: conservatives typically view the world to be a more threatening place than liberals. That, in turn, could influence their position on issues relevant to foreign policy. In addition to being more distrustful of foreigners, they may be more willing to use force. Next to liberals, conservatives certainly are more outspoken in their support of patriotism, a strong military, and the virtue of serving in the armed forces.

Looking at this evidence in total, you'd expect disgust sensitivity to predict voting behavior. Indeed, it does — not perfectly, of course. Your upbringing, religious affiliation, income bracket, and numerous other factors obviously shape your ideology too. But if we look at a large group of people, there's a consistent trend in the data.

In a study of 237 Dutch citizens published in 2014, those who scored highest on an online test of disgust sensitivity were more likely than their less squeamish countrymen to vote for the socially conservative Freedom Party (Partij voor de Vrijheid), which takes a strong position against immigration, is hostile toward Islam, emphasizes the value of Dutch traditions over a multicultural ethos, and is resistant to joining the European Union. The Netherlands has ten political parties whose positions on many issues cannot be neatly characterized as either liberal or conservative, so the researchers could not predict voting across the board, but they did find that subjects' disgust sensitivity correlated with their political ideology in keeping with the pattern previously outlined.

A still larger online study yielded similar findings. Conducted by a team that included Pizarro, Haidt, and Yoel Inbar, it involved twenty-five thousand Americans who were surveyed at the time of the 2008 presidential election. Respondents who scored high on a measure of contagion anxiety were more likely to report that they intended to vote for John McCain (the more conservative candidate) rather than Barack Obama. What's more, a state's average level of contamination concern — calculated from the disgust-sensitivity scores of respondents from each state — predicted the share of the votes actually cast for McCain.

The researchers found the same correlation between disgust sensitivity and political ideology across 122 nations scattered around the globe — basically all countries for which there was a sufficient number of respondents to permit statistical analysis. As the investigators wrote in the *Journal of Social Psychological and Personality Science,* "this strongly suggests" that the "relationship is not a product of the unique characteristics of U.S. (or, more broadly, Western democratic) political systems. Rather, it appears that disgust sensitivity is related to a conservatism across a wide variety of cultures, geographic regions, and political systems."

Not surprisingly, politicians have sought to harness the science of disgust to their own advantage. A noteworthy example is a novel ad campaign launched by candidate Carl Paladino, a Tea Party activist, during the 2010 Republican gubernatorial primary race in New York. Just days before the election, registered voters in his party opened their mailboxes to find brochures saturated with the scent of garbage bearing the message *Something Stinks in Albany*. The mailer displayed photos of Democrats in the state who'd recently been dogged with scandal and characterized Paladino's opponent, former Representative Rick Lazio, as being "liberal" and part of a government that allowed corruption to flourish. Whether the stinky mailers boosted votes for Paladino even slightly we'll never know. But they don't seem to have hurt him. He trounced Lazio by a whopping margin of 24 percent.

More recently, Donald Trump bizarrely characterized Hillary Clinton's extended bathroom break during a Democratic primary debate as "too disgusting" to talk about, causing the crowd to erupt in laughter and applause.

A fear of germs does more than slant people's religious and political views. It literally leads them to think of morality in black-and-white terms — a finding with disturbing ramifications for the criminal justice system. You've likely noted that fairy godmothers always wear white and wicked witches black and, what's more, that the gun-toting heroes and villains in TV Westerns typically follow the same dress code. To Gary D. Sherman and Gerald L. Clore, the psychologists who showed that we associate dark colors with filth and contagion, this seemingly trite observation raised an intriguing question: As a by-product of being honed to spot contaminants, does the human mind actually encode black as sinful and white as virtuous?

To explore this possibility, they took advantage of a favorite brain-training game, the Stroop test. A typical challenge is to press a key the moment you see a word that spells out the name of a specific color — say, yellow. If the letters of the word are colored yellow, people perform the task faster than if the letters are colored blue or another mismatching hue, indicating that the mind takes extra time to process information that conflicts with expectations.

In the researchers' modified version of the test, volunteers were shown morally charged words like *crime, honesty, greed,* and *saint* in randomly alternating black or white type. The words flashed before them at a quick clip and the challenge was to press a key as soon as they recognized them. Subjects were much faster at the task if a word with positive moral connotations was in white or if a word with negative moral connotations was in black, suggesting that connection was rapid and automatic. The opposite pairings evidently triggered confusion, slowing their speed.

In search of better evidence that participants' mental bias might be related to the behavioral immune system, the researchers took the experiment a step further. They primed subjects to think of unethical deeds by having them write about a sleazy lawyer, and then they ran them through the Stroop test again. This time participants were even quicker at linking black words to evil and white words to virtue — and this despite the fact that some of the words used during the trial, among them, *gossip, duty,* and *helping,* were much more loosely related to morality. Since the behavioral immune system operates at high speed to protect us from germs — indeed, some scientists liken it to a reflex — the researchers grew increasingly confident that the subjects were relying on moral intuition rather than the slower process of conscious reasoning.

If so, Sherman and Clore theorized, people who were the fastest to link white to morality and black to immorality would be more concerned about germs and cleanliness. To explore this hunch, all the participants were asked to evaluate the desirability of cleaning products and other consumer goods at the end of the trial. Just as they an-

ticipated, those whose test results suggested they might be germaphobic gave the most favorable ratings to cleaning products — especially
items that were hygiene-related, like soap and toothpaste.

Since the tendency to see black as bad is heightened when moral
issues are foremost on our minds, a courtroom is exactly the place
where one would expect the cognitive bias to be most pronounced —
unsettling news for people of color hoping for a fair trial. "The darkness-contamination-evil link probably doesn't contribute as strongly
to prejudice as the linking of ethnicity, poverty, and crime," said Clore,
"but it's concerning because all these negative biases might have an
additive effect, raising the odds that a person of color will be found
guilty or receive a harsher sentence." (Sherman and Clore's trial was
not designed to test if subjects' own skin color influenced how inclined
they were to link dark shades to evil, so whether people of different
ethnicities are equally prone to the same bias remains to be seen.)

These and related studies raise an obvious question: How have
parasites managed to insinuate themselves into our moral code? The
wiring scheme of the brain, some scientists believe, holds the key to
this mystery. Visceral disgust — that part of you that wants to scream
"Yuck!" when you see an overflowing toilet or think about eating cockroaches — typically engages the anterior insula, an ancient part of the
brain that governs the vomiting response. Yet the very same part of the
brain also fires up in revulsion when subjects are outraged by the cruel
or unjust treatment of others. That's not to say that visceral and moral
disgust perfectly overlap in the brain, but they use enough of the same
circuitry that the feelings they evoke can sometimes bleed together,
warping judgment.

While there are shortcomings in the design of the neural hardware
that supports our moral sentiments, there's still much to admire about
it. In one notable study by a group of psychiatrists and political scientists led by Christopher T. Dawes, subjects had their brains imaged as
they played games that required them to divide monetary gains among
the group. The anterior insula was activated when a participant decided to forfeit his own earnings so as to reallocate money from play-

ers with the highest income to those with the lowest (a phenomenon aptly dubbed the Robin Hood impulse). The anterior insula, other research has shown, also glows bright when a player feels that he has been made an unfair offer during an ultimatum game. In addition, it's activated when a person chooses to punish selfish or greedy players.

These kinds of studies have led neuroscientists to characterize the anterior insula as a fountainhead of prosocial emotions. It is credited for giving rise to compassion, generosity, and reciprocity or, if an individual harms others, remorse, shame, and atonement. By no means, however, is the insula the only neural area involved in processing both visceral and moral disgust. Some scientists think the greatest overlap in the two types of revulsion may occur in the amygdala, another ancient part of the brain.

Psychopaths — whose ranks swell with remorseless cold-blooded killers — are notorious for their lack of empathy, and they typically have smaller than normal amygdalae *and* insulae, along with other areas involved in the processing of emotion. Psychopaths are also less bothered than most people by foul odors, feces, and bodily fluids, tolerating them — as one scientific article put it — "with equanimity."

People with Huntington's disease — a hereditary disorder that causes neurological degeneration — are similar to psychopaths in having shrunken insulae. And they, too, lack empathy, though they don't exhibit the same predatory behavior. Possibly owing to damage to additional circuits involved in disgust, however, the afflicted are remarkable in showing no aversion whatsoever to contaminants — for example, they think nothing of picking feces up with their bare hands.

Interestingly, women rarely become psychopaths — the disorder affects ten males for every one female — and they have larger insulae than men relative to total brain size. This anatomical distinction may explain why they're the sex most sensitive to disgust, and it may also have bearing on yet another traditionally feminine characteristic: as befits women's role as primary caretakers, they score higher than men on tests of empathy — a very useful trait for gauging when a cranky baby has a fever or needs a nap.

Why moral and visceral disgust became entangled in our brains in the first place is harder to explain, but British disgustologist Valerie Curtis puts forward a scenario that, while impossible to verify, certainly sounds plausible. Evidence from prehistoric campsites, she notes, suggests that our ancient ancestors may have been more concerned about hygiene and sanitation than commonly assumed. Some of the earliest artifacts from these sites include combs and middens (designated dumpsites for animal bones, shells, plant remnants, human excrement, and other waste that might attract vermin or predators). Early humans, she strongly suspects, would have taken a dim view of peers who were slobs about disposing of their garbage, spat or defecated wherever they pleased, or made no effort to comb the lice out of their hair. These inconsiderate acts, which exposed the group to bad odors, bodily waste, and infection, triggered revulsion, and so, by association, the offenders themselves became disgusting. To bring their behavior into line, Curtis thinks, they were shamed and ostracized, and if that failed, they were shunned—which is exactly how we react to contaminants. We want nothing to do with them.

Since similar responses were required to counter both types of threat, the neural circuitry that evolved to limit exposure to parasites could easily be adapted to serve the broader function of avoiding people whose behavior endangered health. Complementing this view, Curtis's team found that people who are the most repulsed by unhygienic behavior score higher than average on a test of orientation toward punishment—that is, they are the most likely to endorse throwing criminals into jail and imposing stiff penalties on those who break society's rules.

From this point in human social development, it took just a tad more rejiggering of the same circuitry to bring our species to a momentous place: We became disgusted by people who behaved immorally. This development, Curtis argues, is central to understanding how we became an extraordinarily social and cooperative species, capable of putting our minds together to solve problems, create new inventions, exploit natural resources with unprecedented efficiency, and ultimately lay the foundations for civilization.

"Look around you," she said. "There's not one single thing in your life that you could have made on your own. The massive division of labor [in modern societies] has incredibly increased productivity. The energy throughput of humans nowadays is a hundred times what it was in hunter-gatherer times." The big question is: "How have we done this clever trick? How are we able to work together?"

Explaining why we might be induced to cooperate is not an easy task. Indeed, it has stymied many an evolutionary theorist. The gist of the problem is this: By nature, we are not altruists. When you bring people into a lab and have them play games with different rules to earn money, there are always the greedy who don't mind if other people go away empty-handed. There are always those who will cheat if they think that they can get away with it. From endless iterations of these experiments, this much is clear: People cooperate only if it is more expensive for them *not* to cooperate. The selfish must be punished.

Today, we have laws and police officers to enforce them. But they are modern inventions, and they build on something far more fundamental, the glue that has always held society together. Indeed, society would not exist were it not for this cohesive force — namely, disgust.

"If you're greedy, if you cheat me or steal my things, I could beat you up," Curtis said. "But you might hit me back. And you might get your big, strong brothers who will beat me up as well. So it's probably not a great thing to do. It would be much better if I say, 'She's disgusting, she's behaving like a social parasite, she's taking more than her fair share of the cake,' and to shun you. So I'm using my mental equipment that I've got from disgust to punish you. I'm punishing you by exclusion, not by a violent act. It doesn't cost me anything. It's hard for you to get back at me. And I can also get my big brother and talk to him and say, 'Do you know what she did? She is so disgusting.' And he'll go, 'Oh yeah, she's disgusting,'" and spread the word.

Darwin thought our species' social values might be driven by an obsession with "the praise and blame of our fellow man." Indeed, we care more about our reputations than whether we're really in the right or not. The face of contempt, which, Darwin noted, is identical to that of disgust, is a powerful deterrent. In prehistoric times, being excluded from the group for antisocial behavior would have been tantamount to a death sentence. It is very hard to survive in the wild by your own skills, fortitude, and wit alone. Natural selection would have favored cooperators, people who played by the rules and reciprocated in kind.

Disgust's use to curb the behavior of the inconsiderate and selfish — including people whose poor hygiene threatened the welfare of the group — was essential for our ancestors' technological advancement in another way as well. While sociality offers extraordinary benefits — we can trade goods, exchange labor, forge new alliances, and combine ideas — it also comes at a steep price. We are walking bags of germs. Working in proximity to others exposes everyone to infection and illness. To gain the benefits of cooperation without this huge risk, we must "do this dance," said Curtis. By that she means that we must get close enough to collaborate but not so close as to jeopardize our health. We humans needed rules for achieving this delicate balancing act, and so we acquired manners.

"From a very early age, we learn to be continent with our bodily fluids, not making nasty smells, not eating with an open mouth or spitting. It's highly adaptive because that means you can sustain a social life at a lower [health] cost. People who break these rules are very rapidly socially excluded," Curtis said.

To her, manners are what separated us from animals and allowed us to take "the first baby steps" en route to becoming civilized supercooperators. Indeed, she thinks manners may have paved the way for "the great leap forward," an explosion of creativity fifty thousand years ago manifested by specialized hunting tools, jewelry, cave paintings, and other innovations — the first signs that humans were sharing knowledge and skills and working together productively.

Manners set our species on the track toward progress, but to truly become civilized, humans needed a more elaborate code of conduct, one that would bind communities together. They needed religion. Fortunately for humanity, it emerged just when it was most needed — namely, when our ancestors stopped wandering the world and decided to put down literal roots.

About ten thousand years ago, some hunter-gatherers began experimenting with a radical new lifestyle: farming. Only a few of them did so at first, but the movement picked up steam, and gradually more people began to settle down, trading the wandering life for a patch of land, typically by a river delta.

Infectious diseases spread with alarming efficiency when there are a large number of hosts living near one another, especially under unsanitary circumstances. The advance of agriculture created those very conditions.

The first farmers could barely eke out an existence, being one crop failure away from disaster. Their grain-heavy diet was deficient in many nutrients and overabundant in others (the bacteria that cause cavities thrived on all those carbohydrates, triggering dental woes unknown to hunter-gatherers). Hunger and malnourishment combined to weaken their immune systems, making them more vulnerable to infection.

As they became more successful at farming, ironically, their health problems only worsened. Their grain stores attracted insects and vermin that spread disease. With human settlement came piles of human waste and a greater danger that the water people drank was polluted with fecal contaminants. And the chickens, pigs, and other animals that they domesticated brought them in contact with new pathogens for which they had no natural resistance.

As these risks mounted, early farmers fell prey to wave upon wave of diseases — many unheard-of in prehistoric times — including mumps, influenza, smallpox, whooping cough, measles, and dysentery, to mention just a few.

This didn't happen overnight. It took thousands of years for agriculture to take off. Few cities in the Middle East, where the movement began, had more than fifty thousand inhabitants prior to biblical times. So the perfect storm was slow to gather, but when it hit, a health crisis of unimaginable disruption and trauma ensued. These new diseases were far more lethal and terrifying than the versions manifested in the untreated and unvaccinated today. We are the heirs of exceptionally hardy people who were unusual in having immune systems that could repel these virulent germs. Those at the forefront of these epidemics likely fared far worse on average than our more recent ancestors. Consider the fate that awaited some of the first people to get syphilis: Pustules popped up on their skin from their heads to their knees, then their flesh began to fall off their bodies, and within three months they were dead. Those lucky enough to survive the ravages of never-before-encountered germs rarely came away unscathed. Many were crippled, paralyzed, disfigured, blinded, or otherwise maimed.

It was exactly at this critical juncture that our forefathers went from being not particularly spiritual to embracing religion — and not just passing fads, but some of the most widely followed faiths in the world today, faiths whose gods promised to reward the good and punish the evil. (Hunter-gatherers, at least today, sometimes believe that spirits can influence the weather or other events, but these mystical beings are rarely concerned about whether humans behave in a moral fashion.) One of the oldest of these enduring belief systems is Judaism, whose most hallowed prophet, Moses, is equally revered in Christianity and in Islam (in the Koran, he goes by the name Musa and is referred to more times than Muhammad). Half the world's population follows religions derived from Mosaic Law — that is, God's commandments as communicated to Moses.

Not surprisingly, given its vintage, Mosaic Law is obsessed with matters related to cleanliness and lifestyle factors that we now know play a key role in the spread of disease. Just as villages in the Fertile Crescent were giving rise to filthy, crowded cities, and outbreaks of illness were becoming an everyday horror, Mosaic Law decreed that Jewish priests should wash their hands — to this day, one of the most effective public-health measures known to science.

The Torah contains much more medical wisdom — and by that I don't mean merely its famous admonishments to avoid eating pork (a source of trichinosis, a parasitic disease caused by a roundworm) and shellfish (filter feeders that concentrate contaminants) and to circumcise sons (bacteria can collect under the foreskin flap, so removing it is believed to have lowered the spread of STDs).

Jews were instructed to bathe on the Sabbath (every Saturday); cover their wells (a good idea, as it kept out vermin and insects); engage in cleansing rituals if exposed to bodily fluids like blood, feces, pus, and semen; quarantine people with leprosy and other skin diseases and, if infection persisted in the community, burn their clothes; bury the dead quickly before corpses decomposed; submerge dishes and eating utensils in boiling water after use; never consume the flesh of an animal that had died of natural causes (a sign that it might have been felled by illness) or eat meat more than two days old (likely on the verge of turning rancid).

When it came time for divvying up the spoils of war, Jewish doctrine required any metal booty that could withstand intense heat — objects made of gold, silver, bronze, or tin — to "be put through fire" (sterilized by high temperatures). What could not endure fire was to be washed with "purifying water": a mixture of water, ash, and animal fat, basically an early recipe for soap.

Equally prescient from the standpoint of modern disease control, Mosaic Law has numerous injunctions specifically related to sex. Parents were admonished not to allow their daughters to become prostitutes, and premarital sex, adultery, male homosexuality, and bestiality were all discouraged, if not banned outright.

Religion is an ideal enforcer of good public health, for many of the behaviors most relevant to disease propagation occur behind closed doors, outside of public view. There's simply no way of getting around an omnipresent, all-seeing God ever on the lookout for those who defy His will. Lest His flock be tempted to stray from the fold, the Torah makes clear that there will be a steep health cost. The Lord, it warns, will punish the disobedient with "severe burning fever," "the boils of

Egypt," "with the scab, and with the itch," "with madness and blindness" — and, if all that fails, the sword.

To quote John Durant, author of the *Paleo Manifesto,* a book about
ancient health wisdom and the basis for the above summary of the
medical sophistication of the Torah:

> Taken as a whole, the knowledge of hygiene contained in the Mosaic
> Law is nothing short of stunning. It correctly identifies the main
> sources of infection as vermin, insects, corpses, bodily fluids, food
> (especially meat), sexual behaviors, sick people, and other contami-
> nated people or things. It implies that the underlying source of in-
> fection is usually invisible and can spread by the slightest physical
> contact . . . And it prescribes effective methods of disinfection, such
> as hand washing, bathing, sterilization by fire, boiling, soap, quaran-
> tine, hair removal, and even nail care.

Needless to say, the axiom "Cleanliness is next to godliness" originated in Mosaic Law — and hence was subsequently embraced by
Christianity and Islam. But Hinduism, which evolved more independently, is equally obsessed with bathing before prayer and concerned
about contamination of the body and what parts of it should be allowed to touch other objects or people (the left hand, for example,
is used strictly for bathroom functions, so it is a grave offense for a
Hindu to offer food to someone with that hand).

Of course the world's great religions are about much more than hygiene. Indeed, they tend to be most preoccupied by matters related to
spiritual purity, sacred duty, and the preservation of the soul. But the
use of disgust to punish people whose health practices endangered the
group could easily be leveraged for the purpose of drumming up moral
outrage to condemn the cruel, the greedy, and the malevolent. This
repurposing of the emotion gave society two benefits for the price of
one, for antisocial behavior, like hygiene infractions, would be hard to
police without the deterrent of an all-knowing God with a vengeful
streak.

We may owe disgust a great debt for our manners, morals, and religion — and ultimately our laws, politics, and government, as the latter three can be built only on the former. Evolution got the ball rolling by making our ancestors revolted by parasites and any behavior that exposed them to infection, then culture took over and transformed people into super-cooperators willing to abide by shared codes of conduct. At least, that's one version of how, over the eons, scattered tribes of nomads united to become global citizens whose minds are now wired together by the Internet.

I find this perspective on human history compelling in broad strokes save for one caveat: it may shortchange biology's role in our species' recent moral development. Contrary to common assumption, human brains didn't stop changing once people submitted to divine authority and became civilized. They kept on changing — especially, perhaps, in the very regions involved in processing disgust.

Admittedly, that's conjecture. But discoveries from the forefront of genetics support my thinking. One of the most surprising findings to emerge from human gene-sequencing data in the past decade is that human evolution has been speeding up in recent times. In fact, adaptive mutations in our species' genome have accumulated a hundred times more quickly since farming got under way than at any other period in human history, and the closer we move to the present, the quicker the adaptive mutations pile up.

Scientists were initially puzzled by this unexpected finding until it finally dawned on them that the catalyst behind this change was ourselves. Humans were radically transforming their environment by taking up the plow, and their bodies and behavior had to adjust to the rapidly shifting landscape. In an evolutionary eyeblink, they had to adapt to new diets and very different lifestyles. Our species' cooperative spirit — our ingenuity and ability to work together — forced us into evolution's fast lane.

The quickest-changing sections of the human genome regulate the functioning of the immune system and the brain. Given disgust's role in coordinating our physical and behavioral defenses against infection, it stands to reason that the parts of the brain that the emotion engages could have undergone significant remodeling with the rise of civilization.

That argument is even more convincing when you consider that large segments of populations were decimated by plague and pestilence over that very period. Natural selection would have strongly favored people who believed in God or who, at the very least, were conscientious in obeying religious doctrine that served to protect their health. Most important, it would have favored the survival of people with a punitive streak — that is, those prone to stiffly penalize anyone who broke society's rules. And as agriculture gave way to industry, causing a massive migration from farms to factories and concentrating more people than ever before into sprawling squalid slums, these pressures surely would only have intensified.

WHILE THERE IS UNCERTAINTY as to when and how disgust became embedded in our system of ethics, there can be no doubt that its influence on society has been transformative. Without this powerful emotion to keep us all in line, we could not have achieved as much as a species. Miraculously, disgust has gotten us to cooperate without raising a fist — indeed, often without so much as a slap on the wrist. It has elicited so much good merely by the shaming and shunning of those whose actions harm the group.

For that reason, some thinkers have come to view disgust as God's gift to us. Leon Kass, chairman of the President's Council on Bioethics under the administration of George W. Bush, counseled that we should heed "the wisdom of repugnance." This voice that wells up inside us warns when a moral boundary has been crossed, he argued. In an article for the *New Republic,* he called for people to listen to its outrage at acts like human cloning, abortion, incest, and bestiality. Repugnance, he wrote, "speaks up to defend the central core of our humanity. Shallow are the souls who have forgotten how to shudder."

Needless to say, Pizarro has a less rosy view of disgust — and not without cause. As we've seen, it can make prejudice *feel* right, justifying the stigmatization of immigrants, homosexuals, the homeless, the obese, and other vulnerable groups. Moreover, our natural revulsion to disease has fed into the notion that sickness is God's punishment for sin — a view that still persists around the world even as modern medicine has dramatically advanced.

Our brains are also prone to viewing primary disgust elicitors like blood and semen as agents of evil. In many cultures, a woman who has been raped is treated like a sinner. She is tarnished, sullied, no longer virtuous or to be valued. No man will be with her because she has been corrupted by another man's crime. The fact that women menstruate has further stoked the flames of misogyny, for this "bad blood" is often seen as God's curse, proof of their inferior moral status. In many cultures, menstruating women are confined to separate quarters so as not to contaminate others. Orthodox Jews are prohibited from sitting on a chair that a menstruating woman has occupied. Hindus must bathe and change their clothes if they come in contact with women in this "impure" state. Even in more secular pockets of the world, many couples — both men and women — believe it is wrong to have sex when a woman has her period. Owing to how disgust affects our thinking, it's all too easy for women to be viewed as both polluting and morally offensive and thus deserving of fewer rights than men.

From a legal standpoint, disgust is also problematic — and not just because of the racist implications of having minds that equate dark skin with contamination and sin. Disgust leads us to view gory crimes as the most egregious of offenses and therefore worthy of the severest punishment. Consequently, the murderer who cuts a person's throat is likely to get a stiffer sentence than the one who kills more tastefully — say, by adding a dash of arsenic to the victim's tea or pressing a pillow over his face. Admittedly, a dead body is still not pretty, but an intact corpse tends to go over better with juries than one smeared with blood and chopped into pieces.

Pizarro is disturbed by the logic of putting someone away longer for a crime that is grisly versus one that's cleanly executed. "It's a tricky question," he said. "Do you show the gory pictures of the murder during sentencing?" As he points out, those images have nothing to do with whether or not the defendant committed the crime. What's more, he said, "The judge can't just say, 'Don't let this emotion influence you.' It would be great if that's how human beings work, but you can't undo that."

Even more troublesome, a study of people serving as mock jurors found that those highly prone to disgust were more inclined to judge ambiguous evidence as proof of criminal wrongdoing, to impose stiffer sentences, and to see the suspect as wicked. Compared to their less easily revolted counterparts, they were also more prone to harboring an exaggerated sense of the prevalence of crime in their own neighborhoods. A related study whose participants included law students, police cadets, and forensic experts similarly showed that disgust sensitivity correlated with a tendency to judge crime more severely and punish the perpetrators with longer sentences — and this association held up even for veteran forensic experts who were accustomed to seeing gruesome evidence. To put it plainly, prosecutors benefit from having jurors with acute sensitivity to disgust, while defense attorneys (and the defendants) gain from having jurors with the reverse disposition.

"I've been approached by people who do work for jury selection," said Pizarro, "and they wanted to know what to tell lawyers about this. It creeped me out because you really could use this emotion to your advantage, and I don't want to be a part of that."

Of course, if disgust makes us less tolerant of people who break the law, then arguably we should welcome it into our lives. Pizarro is not persuaded by that line of logic. "I do a podcast with a philosopher friend of mine and his take on this is 'To the extent that disgust can fuel your belief that, say, molesting children is wrong, then bring on disgust.' My response is 'I hope that you'd be opposed to child molestation on plenty of other grounds that don't require being grossed out at the thought of it.'" Still, he conceded, "Maybe this is harder to do in real life."

While people may not be able to suppress their moral intuitions, Pizarro would like them to challenge these sentiments with reason and logic. It may take long and arduous intellectual work to reach an ethical decision — for example, that slavery should be abolished or it's cruel to eat animals — but with the passage of time, he said, our new values can become automatic and intuitive.

If more people favored reason over emotion in making moral decisions, would politics be less polarized?

"We think of ethical views as wildly different across individuals and across cultures, but the truth is that there's a ton of agreement," said Pizarro. "Most people think that murder, rape, stealing, lying, and cheating are wrong. What's interesting is where they diverge. Those differences have become a hotbed of political rhetoric and abuse." And as he notes, where people clash predominantly relates to sexual mores and other social values highly pertinent to the transmission of disease. Which potentially implies something radical: it's parasites that have divided us! So if we could eradicate the worst of them and tamp down our disgust, perhaps people's attitudes would change and political debates would not be so rancorous.

Of course, that's absurdly simplistic. Abortion may become more heinous if you're easily disgusted, but fundamentally this lightning-rod issue is about whether or not you think it's murder. Opposition to gay rights may stem from the belief that children will be better off in a traditional family, one headed by husband and wife, rather than revulsion at the thought of anal sex. Hostility toward immigrants is largely based on concerns that they're taking jobs away from a nation's citizens or might pose a security threat — not fears that they're going to make people sick. Not everything is about parasites!

With that caveat in mind, I invite you to entertain an even wilder idea in the next chapter. Maybe we've *underestimated* parasites' political clout. Maybe they permeate our entire worldview. Maybe geopolitics should be taught from a parasite's point of view.

12

The Geography of Thought

D O YOU PUT THE WELL-BEING of your community before your own happiness?

For decades, sociologists have been puzzled by a marked contrast in how different parts of the world answer that question. People living in North America and Europe are more likely to see themselves as free agents responsible for their own personal happiness and success, a mind-set epitomized by the western United States' reverence for rugged individualism. In contrast, large swaths of the East — notably India, Pakistan, and China — place high value on collectivism, and group cohesion and harmony take precedence over individual aspirations. This attitude is by no means limited to Asia, however. Equatorial regions of South America and Africa rank highest of all on a standardized questionnaire measuring collectivist attitudes. Where a society falls on the collectivism-versus-individualism spectrum in turn correlates with a host of other traits, from religious values and political views to attitudes toward strangers.

In 2007, Randy Thornhill, a biologist at the University of New Mexico, and his graduate student Corey Fincher pondered the origins of this powerful cultural divide. Over the years, social scientists had put forward a hodgepodge of theories to explain it, many focusing on historical idiosyncrasies, economic development, and means of livelihood. But the biologists found these explanations lacking. None accounted for why the myriad traits defining each cultural perspective clustered together in the first place.

That got them thinking, Fincher said. "Can we explain personalities and attitudes based on natural selection or evolutionary history? What aspects of an environment might promote different personalities?" The concentration of collectivist cultures near the equator, a parasitic hot zone, caught their attention. Our vulnerability to infection, they observed, shapes a number of aspects of culture — how spicy people like their food, for example, and the degree to which they prize a mate's beauty, which in large part is a marker of a strong immune system. If parasites can influence our customs and aesthetics, they reasoned, perhaps they can shape our temperament and values as well.

As they prepared to test their hunch, unbeknownst to them, another group of investigators was proceeding on a parallel front, prompted by a different line of reasoning. That team was headed by someone we've already met — Canadian psychologist Mark Schaller, who'd found that people become more prejudiced toward foreigners when shown pictures that called to mind the threat of infectious disease. Other psychologists, following his lead, had shown that subjects become more reticent and less sexually adventurous when primed to think of germs. Inspired by these observations, Schaller and colleague Damian Murray set out to explore whether these brief lab-induced shifts in disposition might become long-lasting traits in places where parasitic illness was an everyday threat. To help them in that task, they were able to tap into two newly available databases created by large cadres of scientists working for the International Sexuality Description Project and a separate initiative, the Personality Profiles of Cultures Project.

By superimposing this data onto old disease atlases, Schaller and Murray uncovered a fascinating pattern, one that dovetailed beauti-

fully with experimental findings. In areas that have historically endured a high incidence of infection, people are more likely to be introverted and less prone to seek out novel experiences. Women and, to a
lesser extent, men in these regions also report having restricted sexual
lifestyles — that is, they have fewer partners over their lifespans and
believe sex should be reserved for stable, committed relationships.
In short, they don't mingle readily and they tend to follow traditional
codes of conduct that might serve as a buffer against disease — for example, washing before prayer, bowing instead of shaking hands, and
marrying only within their religious groups.

Their traits, Schaller and Murray realized, seemed to be part of
a broader package of values — namely, collectivism. An adherence
to convention and distrust of foreign ways of doing things are well-
known characteristics of collectivist societies. Could disease avoidance be an overlooked function of that belief system?

As they turned to test that theory, they received a call from Thornhill, who'd just heard about their work from a colleague who'd attended a recent conference where they'd presented their findings.
Thornhill and his graduate student were doing the same kind of research and thinking along very similar lines. Would they like to join
forces?

"I was shocked," said Schaller. "This is messy stuff to do and also
kind of wacky. Yet here was a group working essentially in the same
territory." Still more remarkable, their approaches appeared to be
highly compatible. "They were coming from the perspective of ecolo-
gists and they knew a lot about the spread of diseases. We were approaching this topic from the perspective of psychologists who knew
how pathogens affect behavior." What's more, each group had developed complementary measures for assessing the prevalence of infectious disease. Thornhill and Fincher had compiled data on ninety-
eight countries from contemporary sources such as the Global
Infectious Diseases and Epidemiology Network. Schaller and Murray
had combed through old medical atlases to assess where infectious
disease had been highest in the past. If they collaborated, they would

be able to test their theory on both data sets — a big bonus, as historic levels of disease should more strongly correlate with collectivism than present-day figures, assuming a region's health was driving its value system, as they were proposing.

The results suggested their "wacky" idea might not be so wacky after all. High levels of infectious disease were a remarkably powerful predictor of collectivism. And, just as they'd theorized, the association was strongest of all for regions that had historically been burdened by heavy parasite loads. The association held up even after they statistically controlled for variables such as poverty, population density, and life expectancy. "We're not saying," cautioned Fincher, "every individual in the U.S. is individualistic or everyone in China is collectivistic. There's tremendous variation within any given population on these traits. We're just talking about relative prevalence — mean or average values — where you'll see differences between nations."

The researchers dubbed their theory the "parasite-stress model of sociality" and two of the scientists — Fincher and Thornhill — soon went on to examine whether the pattern they'd detected on an international scale held up within the United States. Sure enough, they found that Americans were the most collectivistic in the very states — mostly in the Deep South — where Centers for Disease Control and Prevention figures indicated infectious disease was highest.

Thornhill, who grew up in Alabama in the forties, was not surprised to learn southerners had unusually tight family bonds and made a big distinction between "them" and "us." In the early decades of the twentieth century, southerners were considered far less productive than northerners, a sluggishness that was eventually traced to epidemic levels of hookworm that had rendered a large sector of the population anemic. The South in that era was also plagued by malaria, a problem that required the draining of swamps and other major efforts to bring it under control. He believes those combined factors may explain why the region even today remains more clannish than the rest of the country. The vestiges of this ethnocentrism, he said, is even reflected in the way people speak. *"Y'all* was invented in the

American South in the eighteenth century. It's not a contraction of *you* and *all*." Today it may be used that way, but it was originally a linguistic tool for addressing the person's in-group — extended families and close friends. "When languages drop [first- and second-person] pronouns — *I* and *you* — they're likely to be collectivist cultures. Individualist cultures talk about *I* a lot. The measure is crude but it's considered valid across countries of the world."

The notion that a single phenomenon — our species' psychological adaptation to parasite stress — can shape entire cultures may sound simplistic, but plenty of scientists are open to the theory. Cornell psychologist David Pizarro counts among them. "I love their work," he said. "I think this is the right way of approaching the topic." Steven Pinker, the evolutionary psychologist and author of such best-selling books as *The Blank Slate* and *The Better Angels of Our Nature,* echoed his sentiment. "I think the theory is well worth pursuing."

Of course, it has attracted flak too, the most common criticism being that a correlation between parasite stress and collectivism does not prove that one caused the other. Especially with something as complex and multifaceted as culture, it can be very tricky to control for unrecognized factors that might be behind the association.

The theory's progenitors are well aware of that hazard, and though there's no simple solution to the problem, they have found ways to subject their ideas to more rigorous scrutiny. Based on their model, for example, they've generated numerous new predictions and tested them against voluminous bodies of data gathered from many different sectors of the social sciences. So far, they report, their theory has held up well to this barrage of tests. Not only that, but their latest findings lead them to think its explanatory powers are considerably broader than they'd initially proposed.

Do you live in a democracy or under a brutal dictatorship? Are you deeply religious? Do women in your country have the same rights as men? Is war frequently erupting around you? Thornhill and Fincher believe parasite stress has direct bearing on all those questions.

The scientists have buttressed their theory not just by crunching numbers but also by reviewing findings from field studies. These studies show that parasites are finicky creatures, rather like orchids in a hothouse. Especially in the tropics, they thrive best within narrow ranges of temperature and humidity. Scattered in pockets across Peru and Bolivia, for example, are 124 genetically distinct strains of the human parasite *Leishmania braziliensis*. People in those regions are well adapted to coexisting with only some of those strains. So if they move very far away from their own locale, they may encounter novel strains that could sicken or kill them. If a foreigner were to insinuate himself into their group, his germs could be lethal to them and vice versa. That foreigner's genes, should one of them decide to mate with him, would produce children whose immune systems would be less adept at fighting off local scourges.

For these reasons, Thornhill and Fincher theorize, people in parasitic hot zones should be reluctant to marry outside their own community. And they should develop all kinds of idiosyncratic markers of communal identity — unique dialects, religious practices, culinary customs, modes of dress, jewelry, music, and so forth — that allow each group to tell "them" from "us." In short, the theory predicts that regions teeming with parasites should produce homebodies and a Balkanized social landscape — that is, divided by religions and languages, among other social barriers.

Indeed, that's what they found. They also discovered that people in these zones are more fervent about their faith, as measured by the number of times they pray per week, how often they attend services, and many other indices tracked by ethnographers. Atheism, by contrast, flourishes where there are very low parasite loads.

"Religious scholars," said Thornhill, "have been very interested in religious commitment, but they can't predict which countries will be religious based on current paradigms. Their theories are not very sophisticated — for example, you learn your religion from your parents." In contrast, he and Fincher found, epidemiological data is a very good indicator of where religious fervor will burn brightest. And the more baroque a religion's sacred rituals and the greater the demands on its adherents, the better it is at keeping the congregation tightly bound together and separate from the members of other sects and their parasites.

Thornhill and Fincher, convinced their theory might have still broader applicability, reasoned that in parasitic hot zones, pressure to follow sexual and hygiene-related practices might breed intolerance toward those who buck conventions. Traditions are sacred and must be strictly enforced, creating hierarchical societies. People get used to abiding by rules and bowing to authority and they become less accepting of dissent — the very conditions that might be favorable for the establishment of repressive regimes.

To test their hypothesis, the scientists accessed the Democracy Index, the Human Freedom Index, and other publicly available scales that rank countries on a spectrum from democratic to autocratic based on measures such as voter participation, civil liberties, wealth distribution, and gender equality. Again the results supported their theory: Countries under severe parasite stress were more likely to be controlled by dictators; gender inequality was pronounced, and wealth tended to be concentrated in the hands of a small class of elites. In contrast, countries with the least amount of infectious disease had wealth more equitably distributed; their women were on a more equal footing with men, and individual rights were far more extensive. They were overwhelmingly democracies.

If we are to believe Thornhill and Fincher, parasite stress breeds a distrust of outsiders that contributes to homicides, civil strife, and the division of society by race and class. Almost everywhere they looked, they uncovered evidence supporting their theory. "Who would have thought that parasites would relate to collectivism, personality traits, political beliefs, religiosity, and civil strife? If our theory only worked for collectivism and individualism, that would be cool, but it wouldn't suggest that we're onto something big and important."

By scientific standards, Thornhill has a swashbuckling style. The biologist, whose many accomplishments include important work early in his career on the mating behavior of insects, is known as an academic cowboy, someone not afraid to take on controversial subjects. That has occasionally gotten him into trouble. In 2000, he ignited a firestorm with the publication of a book, *A Natural History of Rape,* that he coauthored with the anthropologist Craig T. Palmer. In it, they argued that rape was best understood in an evolutionary context and challenged the notion that the crime was never sexually motivated. That did not go over well with those who believed it to be purely an act of aggression and who took Thornhill and Palmer's position as a justi-fication for an inexcusable crime, as in, "Your Honor, my genes made me do it." That was not what the researchers meant, which they took great pains to explain during appearances on the *Today* show and an array of news programs. But in the ensuing uproar, Thornhill received numerous death threats, and during the worst of the maelstrom, he had to be escorted around his campus by the university police.

The parasite-stress theory has yet to leap into public awareness, so at least for now, disputes about it remain low-key and within aca-demia. "I haven't gotten any death threats against me [for it] yet," he said. But his penchant for making bold statements with few qualifiers suggests he's still game to play the role of provocateur — for better or worse.

Schaller and Murray have adopted a more cautious tone — Schaller speaks of having "creeping confidence" in aspects of the model — and they have focused their research more narrowly in the areas where they have the greatest expertise. As Schaller told me, religion and civil war are outside his bailiwick. While he's intrigued by the many di-rections in which Thornhill and Fincher's work is leading, he has the strongest trust in the data he and Murray have been most engaged in collecting and analyzing, which pertains to authoritarianism.

They've investigated its link to parasite stress using a different ap-proach than Thornhill and Fincher. Instead of doing a country-by-country analysis, they focused on ninety small-scale societies that are culturally very distinct to guard against the potential bias of regions having a shared history. For example, nations may be democracies not because they have a low incidence of infectious disease but because their populaces are largely derived from Europe and they've imported its system of values. Would pathogen prevalence still predict autocratic governance among societies with almost no shared history? The answer, they found, is yes. They also conducted another study that relied on a more concrete measure of authoritarianism than a paper test. They looked at how many people in a society are right-handed, their logic being that cultures intolerant of individuality would pressure people who are naturally left-handed to use their right hands. If high rates of infection promote authoritarian values, they reasoned, parasitic hot zones should have more right-handers. That's exactly what they found.

"Value systems may have costs and benefits that vary depending on the prevalence of pathogens," said Schaller. "The cost of being a maverick—what we in the West call being a rugged individualist—might outweigh benefits in places where pathogens are abundant, whereas the reverse might hold true where pathogens are scarce." In that world, he said, doing things differently, thinking outside the box, is more likely to be valued, as it spurs creativity and technological innovation.

Although there are now—in Schaller's words—"a lot of different pieces of evidence linking pathogen prevalence to various kinds of cultural differences," he cautioned that parasite stress clearly is not "the only thing that plays a role [in shaping society]. If there's one thing we know in cognitive behavioral sciences, it's this: Everything is multi-determined."

In his view, for example, it would be grossly simplistic to conclude that religion is simply a parasite defense. Even if that's a function it serves—and that's a big *if* for now, he emphasized—that in no way implies that religion arose to serve that purpose alone or that it continues to exist for that one reason.

Similarly, linguistic diversity, types of governance, and outbreaks of violence are no doubt the products of numerous geographic and historical factors, not parasite stress alone. (For an in-depth analysis of such factors, readers may wish to consult Jared Diamond's Pulitzer Prize–winning book, *Guns, Germs and Steel*.) The big challenge, said Schaller, will be to figure out how all these variables combine, which he suspects could take many years to work out.

How parasite stress gets translated into attitudes and personality traits will also need to be clarified. Schaller speculated that "there may have been differential selection for different personality traits in different cultures. There may be different frequencies of alleles [gene variants] for neurotransmitter pathways related to mood and temperament. This is an area where we want to tread cautiously. My collaborators and I are a bunch of white guys and we're saying Europeans are more open-minded and adventurous. It sounds so horribly self-serving. We know how incredibly complicated the genes and environment interaction is."

For example, he and his collaborators theorize that having one's disgust button pressed often during childhood may up- or down-regulate genes associated with temperament or risk aversion.

Obviously, people learn practices in their culture that guard against infection. Individual life experiences — for instance, witnessing all of one's siblings die of disease in childhood — could also make people hypervigilant about germs.

The causal pathway could be still more complicated. The brain may sense when the immune system is in overdrive due to chronic infections, Fincher speculates, and in response switch an individual's mental outlook into a defensive mode manifested by collectivistic thinking.

Thornhill suspects humans may even have evolved the ability to read levels of antibodies circulating in other people's blood, which would tell them whether those around them were harboring lots of parasites.

Are you talking about a sixth sense? I asked him.

"There could be seven to five hundred senses that allow the brain to detect antibody titers and the duration of immune system activation," he replied. "When you look at a person, you're assessing information about their age, all the hormone markers, the symmetry of the face and movement of the person — all this stuff ties to the health stats of an individual. Body odor might also provide information about a person's immune status. There could be multiple things the brain is reading."

WHILE THE ARCHITECTS of the parasite-stress model wrestle with the mechanisms that might explain their findings, other scientists are scrutinizing fundamental assumptions that underpin it, such as the assertion that bunching up into small insular groups (the technical term for this is *assortative sociality*) helps to thwart the spread of disease. A smattering of animal evidence and computer models by population geneticists support the idea, but scientists caution these models are only as good as the data that's fed into them and are still quite crude. Evolutionary psychologist Dan Fessler points out that human groups almost always trade with one another — something animals don't do — which could undermine claims for the theory. Cultural anthropologists have begun to weigh in with their own opinions, and while facets of the model seem feasible to many of them, they also find flaws in its logic. For example, anthropologist Richard Sosis and colleagues at the University of Connecticut point out that proselytizers regularly come into contact with strangers, and some religious practices — such as ritual bloodletting — promote, rather than impede, the transmission of disease. Although no theory — especially not one that purports to account for cultural variation — can always make spot-on predictions, these counterexamples still raise troubling questions, and if enough of them accumulate, the theory could crumble under their weight.

Its proponents might also be accused of having a political ax to grind, as the model could easily be construed as being anti-religion and an across-the-board indictment of collectivism, which corresponds closely to conservative values. Certainly those who hold views on the right-leaning end of the spectrum might understandably question the researchers' impartiality.

In short, the model is controversial and I suspect will only become more so should the debate over its merits spill into the public arena. Thornhill, for his part, isn't worried about whose feathers he might ruffle or whether the model will stand up to scrutiny. "So far, the evidence that we're wrong is wrong," he said with a good-humored twinkle.

And if you're right, I asked him, how does it change things?

"If we're right, then the implication for saving the world is reducing parasite stress. Focus there. A lot of people would say we've got to build schools. Another group would say we have to build economic institutions in these countries to make them better. Our data and ideas say fundamentally that you work on the non-zoonotic diseases [illnesses transmitted from person to person] and then, in due time, when these kids are raised in better environments in terms of disease, then you'll get more liberal-minded kids. You're going to get economic productivity because there won't be as many social barriers. There's more exchange of ideas. You're going to get interest in education and then you're going to save the world."

When I pressed Thornhill to point to a country that exemplifies his philosophy, he shot back, "The whole of the Western world, because gradually there's been a reduction in infectious disease. In the 1920s, you get chlorinated water. By the 1930s you have food-handling and sanitation laws. These spread very rapidly specifically in the West, but not outside the West. Around the same time you got antibiotics. By 1945, you got fluorinated water and it spread very rapidly. That knocked out all the mouth infections. In 1945, you also got DDT, which knocked out all the insect vectors of disease. Malaria went down the drain and [so did] other human diseases [spread by insects].

"In the 1960s, we got the cultural revolution: civil rights, women's rights, the sexual revolution. All that stuff came out of the cleanup of infectious diseases specifically in the West. Liberal parts of the world. Outside of the West none of these things have occurred."

No doubt many might challenge that provocative slant on history, but if we accept his premise, then we'd be wise to reexamine geopolitical objectives with parasite stress in mind. As Ebola's devastating rampage through West Africa brought into sharp relief, much of the developing world still does not have even the most rudimentary medical infrastructure. Vast populations in parasitic hot zones are without hospitals, doctors, drugs, or surgical equipment. Meanwhile, often directed at the very same underserved regions, wealthy countries are throwing billions of dollars into campaigns aimed at containing wars ignited by ethnic hatred and religious intolerance, not to mention dealing with the refugee crises and other tragedies this violence has spawned. By investing more in health care up front, Thornhill's model predicts, Western countries might not have to spend so much on warfare later — and may ultimately reduce human suffering far more effectively.

A group of ethicists at Oxford University led by Russell Powell agree that the parasite-stress theory could potentially change how wealthier nations make foreign-policy decisions. If the thesis is correct, they write in *Behavioral and Brain Sciences,* "then infectious disease–related interventions are likely to have more far-reaching social, economic, and political implications. It is well known that societal choices can affect susceptibilities to infectious disease, but few have imagined a proximate causal pathway through which infectious disease can shape sociopolitical choices."

AT THE START OF THIS BOOK, I warned that we may have less control over our minds than we think, that we may fancy ourselves in the driver's seat but, unbeknownst to us, an invisible passenger may be steering our choices and behaviors.

Indeed, there may be many invisible passengers vying to steer us — perhaps even all at once. Meanwhile, as we roll along, we see road signs warning of dangers ahead. From one moment to the next, your behavioral immune system is warily assessing others, deciding whether you should be warm and friendly, maybe even have sex, or if you should adopt a chillier stance — interactions that, when repeated over the ages and across the world, may even have shaped the cultures that make up human society.

The mother of all manipulators, the boss of all bosses, is DNA — the most prolific replicator of all. It's infected every creature on the planet, forcing a long parade of hosts to devote their entire lives to furthering its transmission. Certainly, genes can goad members of our own species into doing awfully foolish things, like swooning over people who treat them badly, overeating because they evolved in a world of scarcity, or, especially in an era before birth control, having babies at all the wrong times in life.

This odyssey into the hearts and minds of parasites and their hosts has taken us from genes to geopolitics, quite a leap, but the journey, for all we know, may not end there. Maybe we ourselves are organisms inside some cosmic superbeast. What we call the universe is nothing more than a bubble of flatulence in its monstrous, gurgling gut, and we can no more comprehend its complexity and purpose than *E. coli* could imagine what makes a human tick or fathom the vast expanse of time that separates its lifespan from our own.

Perhaps I am getting a wee bit carried away here. My mind is exploding with thoughts of parasites. For all I know, they may even be ensconced in my brain, quietly nibbling away at my sanity. Nature is full of ghastly and glorious surprises. Not that long ago, many would have scoffed at the notion that microbes in the dark, smelly cavities of the body could influence our behavior. And few could have foreseen that a single-celled parasite would have the wherewithal to trick a rat into courting a cat.

To a man with only a hammer, goes the saying, everything looks like a nail. This book presents an unabashedly parasite-centric view of the world. I have filtered the vast sweep of evolution and history through that narrow lens. Though I've tried to be impartial, I'll admit to being in awe of parasites, so it's possible that I may have on occasion given them more credit than they're due. Certainly they've been badly slighted in the past and their abilities are still grossly underappreciated, especially their remarkable powers of mind control and the multiple pathways by which they influence both human and animal behavior.

This science is still young and fluid. But developments in the field have brought us to an exciting place, for by acknowledging and examining parasites' power over us, we are certain to enhance our own power.

Consider where answers to these questions might take us. If parasites are contributing to mental illness or traffic accidents, how can we evict them from our brains or otherwise thwart them? If microbes in our gut can boost our moods and lower our anxiety levels, how can we better harness them? If our fear of contagion is behind culture wars and even real wars, isn't that important to know? We alone among animals are not driven solely by instinct. We can question how the world works and use that knowledge to create powerful medicines and other wonders. We can question our values, and, if we find them lacking, we can endeavor to swim against the current.

Perhaps prejudice will subside as more folks grow distrustful of their moral intuitions and rely on them less often. Perhaps people will start taking probiotics instead of Prozac. It's hard to predict the future. Only this much is certain: Parasites are woven into our psychology and the very fabric of our being. Indeed, we are more microbe than human. We can be confident that this radical new vision of ourselves will open a world of fresh opportunities.

ACKNOWLEDGMENTS

This book might never have been completed — at least not on schedule — were it not for my kind, energetic, and immensely capable husband, Joshua Cohn. Despite having a demanding career of his own, he critiqued my every draft, improving each version — and there were a lot of them. Did I mention that he mows the lawn and does all the shopping and cooking?

My children, Rachel and Daniel, went from teenagers to adults over the span of time I wrote this book. They were also supportive of the endeavor and never once complained about having a half-present mother who babbled incessantly about parasites.

My sister Gisele McAuliffe, who's much better organized than I'll ever be and a whiz with computers, offered sage counsel about working efficiently and was on call around the clock to solve IT problems. Both she and her husband, David Caleb, very kindly went through the manuscript from beginning to end with a fine-tooth comb, pointing out passages that struck the wrong note or needed more finessing.

I am also blessed with wonderful friends who were enthusiastic supporters of the project from the start. A few deserve special mention.

Phoebe Hoban, a best-selling author, educated me on the intricacies of the publishing industry from A to Z and was a constant reassurance whenever I hit a snag.

When, in a moment of panic, I thought I'd never be able to come up with a satisfying conclusion to the book, a musician friend, Tim Devine, offered great ideas on the spur of the moment, even drafting some of them in scintillating prose that made me jealous of his facility as a writer. I was tempted to cut and paste, but thankfully inspiration did strike, albeit only a moment before the final, *final* deadline.

Wallace Ravven, a fellow science writer, helped me at the eleventh hour as well, drawing my attention to passages that he felt were not nuanced enough — and he was right. Based on his criticism, I made last-minute changes that I suspect may have spared me some bad reviews (or so I pray).

I'm also indebted to my literary agent, Zoë Pagnamenta, who always responded to my queries at lightning speed, read every draft of the manuscript seemingly overnight, and provided thoughtful feedback and guidance on every aspect of the book. (Much thanks to Bob Weil at Norton for steering me toward her.)

I was blessed to be assigned a copyeditor who is also a physician. Tracy Roe was not only attentive to every detail but also flagged difficult-to-follow passages and suggested additions to enhance the content.

My copyeditor thanked me profusely for putting the book's references in the proper format, saving her many hours of work, but the credit actually goes to Lindsay Devine, whom I hired to do the job and who did it much better than I ever could.

My editor and publisher, Eamon Dolan, was a delight to work with in every way. He was cheerful, efficient, decisive, knew exactly what he wanted and articulated it well, and cured me of some bad writing habits — "weak interrogative transitions" and nice-sounding but hollow prose that he called "sausage wrapping." For a long time, they were the bane of my existence, but I'm ever so happy that they never made it into print. Thank you, Eamon!

There are many, many scientists who contributed to this book, so I can't possibly mention them all, but here is a list of some who were exceptionally generous with their time: Lene Aarøe, Shelley Adamo, Martin Blaser, Gerald Clore, Stephen Collins, John Cryan, Valerie Curtis, William Eberhard, Andrew Evans, Daniel Fessler, Corey Fincher, Jaroslav Flegr, Benjamin and Lynette Hart, Celia Holland, Patrick House, Michael Huffman, David Hughes, Clemens Walter Janssen, Kevin Lafferty, Frederic Libersat, Emeran Mayer, Glenn McConkey, Janice Moore, Charles Nunn, Michael Bang Petersen, David Pizarro, Teodor Postolache, Robert Poulin, Nicolas Rode, Paul Rozin, Robert Sapolsky, Mark Schaller, Gary Sherman, Frédéric Thomas, Randy Thornhill, E. Fuller Torrey, Ajai Vyas, Michael Walsh, Joanne Webster, Geraldine Wright, Robert Yolken, and Sera Young.

New York Times science writer Carl Zimmer indirectly contributed to this work by penning *Parasite Rex,* a gem of a book published in 2003 that was my first introduction to parasitic manipulators and is a hard act to follow.

NOTES

INTRODUCTION

page

6   *Eyes are bathed:* Randolph M. Nesse and George C. Williams, *Why We Get Sick: The New Science of Darwinian Medicine* (New York: Vintage, 1994), 38.

      As for any microbes: Michael D. Gershon, *The Second Brain: Your Gut Has a Mind of Its Own* (New York: HarperCollins, 1998), 88.

7   *In medieval times:* M. J. Blaser, "Who Are We? Indigenous Microbes and the Ecology of Human Diseases," *European Molecular Biology Organization Reports* 7, no. 10 (2006): 956.

      Within a few centuries: Jared Diamond, *Guns, Germs, and Steel* (New York: W. W. Norton, 1997), 77–78.

      the 1918 Spanish flu: See https://virus.stanford.edu/uda/.

      Malaria, presently among: Sonia Shah, "The Tenacious Buzz of Malaria," *Wall Street Journal,* July 10, 2010, http://www.wsj.com/articles/SB1000142405274 8704111704575354911834340450.

1. BEFORE PARASITES WERE COOL

9   *The answer to:* Janice Moore, interview by the author, September 1, 2012.

11   *Four decades later:* Janice Moore, interview by the author, Massa Marittima, Italy, March 18, 2012.

12   *Numerous crustaceans:* Moore, interview by the author, October 2011.
     Mammals like ourselves: Moore interview, March 18, 2012.

14   *Especially at the outset:* Moore interview, September 1, 2012.

17   *Her experimental apparatus:* Janice Moore, "Parasites That Change the Be-
     havior of Their Host," *Scientific American* (March 1984): 109–15.

19   *In addition to:* Ibid., 109–11.
     Soon the manipulation hypothesis: Moore interview, September 1, 2012.

21   *Since the eighteenth century:* Moore interview, March 18, 2012.

22   *A colleague who stepped:* Robert Poulin, interview by the author, Massa
     Marittima, Italy, March 18, 2012.

2. HITCHING A RIDE

25   *Frédéric Thomas suspected:* Frédéric Thomas, interview by the author,
     Massa Marittima, Italy, March 19, 2012.

31   *The worm, which is now:* C. Zimmer, "The Guinea Worm: A Fond Obituary," *The
     Loom* (blog), *National Geographic,* January 24, 2013, http://phenomena.national
     geographic.com/2013/01/24/the-guinea-worm-a-fond-obituary/.

32   *The moment the tapeworm:* "Dracunculiasis (Guinea-Worm Disease),"
     World Health Organization, May 2015, http://www.who.int/mediacentre/fact
     sheets/fs359/en/.
     Just twenty years ago: D. G. McNeil Jr., "Another Scourge in His Sights," *New
     York Times,* April 22, 2013.
     with fewer than: M. Doucleff, "Going, Going, Almost Gone: A Worm Verges on
     Extinction," *Goats and Soda* (blog), National Public Radio, July 8, 2014.
     "For my own part": Janice Moore, *Parasites and the Behavior of Animals* (Ox-
     ford: Oxford University Press, 2002), Kindle edition, chapter 3.

33   *As the snail morphs:* M. Simon, "Absurd Creature of the Week: The Para-
     sitic Worm That Turns Snails into Disco Zombies," *Wired,* September 19, 2014,
     http://www.wired.com/2014/09/absurd-creature-of-the-week-disco-worm/.
     However, this may not be: Moore, *Parasites and the Behavior of Animals,* chapter 3.
     A tapeworm that infects brine shrimp: Nicolas Rode, interview by the author,
     April 15, 2015; N. Rode et al., "Why Join Groups? Lessons from Parasite-Ma-
     nipulated Artemia," *Ecology Letters* (2013): 1–3, doi: 10.1111/ele.12074.

34   *Such an intuition:* Kevin Lafferty, interview by the author, July 27 and Au-
     gust 3, 2011.

35   *It seemed logical:* K. Lafferty and A. Kimo Morris, "Altered Behavior of Par-
     asitized Killifish Increases Susceptibility to Predation by Bird Final Hosts,"
     Ecology 77, no. 5 (1996): 1390.

36   *To figure out how the parasite could coax its host:* J. C. Shaw et al., "Parasite
     Manipulation of Brain Monoamines in California Killifish (*Fundulus parvipin-*

nis) by the Trematode Euhaplorchis Californiensis," *Proceedings of the Royal Society B* 276 (2009): 1137, doi:10.1098/rspb.2008.1597.

Lafferty recounted: T. Sato et al., "Nematomorph Parasites Drive Energy Flow Through a Riparian Ecosystem," *Ecology* 92, no. 1 (2011): 201.

37 *To understand how:* C. Zimmer, *Parasite Rex* (New York: Simon and Schuster, 2000), Kindle edition, chapter 4. Also see J. C. Koella, F. L. Sorensen, and R. A. Anderson, "The Malaria Parasite, *Plasmodium falciparum,* Increases the Frequency of Multiple Feeding of Its Mosquito Vector, *Anopheles gambiae,*" *Proceedings of the Royal Society B* 265 (1998): 763–68.

38 *Incidentally, the bacterium:* Koella, Sorensen, and Anderson, "The Malaria Parasite," 763.

Plasmodium has still: Zimmer, *Parasite Rex,* chapter 4.

But a study of Kenyan: R. Lacroix et al., "Malaria Infection Increases Attractiveness of Humans to Mosquitoes," *PLoS Biology* 3, no. 9 (September 2005): 1590–93. Also see R.. C. Smallegange et al., "Malaria Infected Mosquitoes Express Enhanced Attraction to Human Odor," *PLoS One* 8 (2013): e63602, doi: 10.1371/journal.pone.0063602 and L. J. Cator et al., "Alterations in Mosquito Behaviour by Malaria Parasites: Potential Impact on Force of Infection," *Malaria Journal* 13 (May 1, 2014): 164, doi: 10.1186/1475-2875-13-164.

39 *In studies of infected:* B. O'Shea et al., "Enhanced Sandfly Attraction to *Leishmania*-Infected Hosts," *Transactions of the Royal Society of Tropical Medicine and Hygiene* 96 (2002): 117–18.

Interestingly, the mosquito-borne: D. G. McNeil Jr., "A Virus May Make Mosquitoes Even Thirstier for Human Blood," *New York Times,* April 2, 2012.

40 *Malaria, despite mountains of money:* "Ten Facts on Malaria," World Health Organization fact sheet, updated November 2015, http://www.who.int/mediacentre/factsheets/fs094/en.

Dengue fever is soaring: "Dengue and Severe Dengue," World Health Organization fact sheet, updated May 2015, http://www.who.int/mediacentre/factsheets/fs117/en/.

The parasites behind: Leishmaniasis FAQs, U.S. Centers for Disease Control and Prevention, updated January 10, 2013, http://www.cdc.gov/parasites/leishmaniasis/gen_info/faqs.html; Plague FAQs, U.S. Centers for Disease Control and Prevention, http://www.cdc.gov/plague/faq/.

A promising avenue: Mark C. Mescher, interview by the author, June 29, 2014.

The disease is rapidly spreading: X. Martini et al., "Infection of an Insect Vector with a Bacterial Plant Pathogen Increases Its Propensity for Dispersal," *PLoS One* 10, no. 6 (2015): e0129373, doi: 10.1371/journal.pone.0129373.

41 *Arriving in southern Florida:* K. M. Wilmoth, "Citrus Greening Bacterium Changes the Behavior of Bugs to Promote Its Own Spread," press release, University of Florida, July 29, 2015, http://www.newswise.com/articles/view/637908?print-article.

in some parts of the peninsula: J. Ball, "Oranges Bug 'Hacks Insect Behaviour,'" BBC News, July 1, 2015.

More recently: Anthony Keinath, "Citrus Greening Disease in Charleston, Five Years Later," *Post and Courier,* April 20, 2014.

In response to: Ball, "Oranges Bug 'Hacks Insect Behaviour.'"

3. ZOMBIFIED

43 *The eight-legged creature:* William Eberhard, interview by the author, January 31, 2013.

47 *Though dwarfed in stature:* Frederic Libersat, interview by the author, March 20, 2012, and November 5, 2015. For a good overview article, see Frederic Libersat and Ram Gal, "Wasp Voodoo Rituals, Venom-Cocktails, and the Zombification of Cockroach Hosts," *Integrative and Comparative Biology* (2014): 1–14, doi:10.1093/icb/icu006.

49 Sacculina *resembles a:* Jens T. Høeg, interview by the author, November 3, 2015. Also, for an exquisite description of *Sacculina,* see C. Zimmer, *Parasite Rex* (New York: Simon and Schuster, 2000), Kindle edition, chapter 4.

51 *Even when this fungus:* David Hughes, interview by the author, August 9, 2013.

53 *It begins over a decade:* Geraldine Wright, interview by the author, August 10, 2013.

54 *She raised the difficulty:* Ibid.; also G. A. Wright et al., "Caffeine in Floral Nectar Enhances a Pollinator's Memory of Reward," *Science* 339, no. 1202 (2013): 1202–4, doi: 10.1126/science.1228806.

55 *Some research indicates:* D. Borota et al., "Post-Study Caffeine Administration Enhances Memory Consolidation in Humans," *Nature Neuroscience* 17, no. 2 (February 2014): 201–3. Also see I. Sample, "Coffee May Boost Brain's Ability to Store Long-Term Memories, Study Claims," *Guardian,* January 12, 2014, and S. E. Favila and B. A. Kuhl, "Stimulating Memory Consolidation," *Nature Neuroscience* 17, no. 2 (February 2014): 151–52.

Intriguingly, the drug: Michael Yassa, interview by the author, November 4, 2015.

"It's pretty funny": Wright interview.

4. HYPNOTIZED

57 *He believed his mind:* Jaroslav Flegr, interview by the author, summer 2011 and September 21 and September 22, 2011.

58 *I'd been pointed:* Robert Sapolsky, interview by the author, summer 2011.

Medical evidence also: E. Fuller Torrey, interview by the author, summer 2011.

59   *Joanne Webster, a parasitologist:* Joanne Webster, interview by the author, summer 2011.

At the Stanley Medical Research: Torrey interview.

It has been a long uphill battle: Flegr interview.

61   *Delving deeper into:* W. M. Hutchison, P. P. Aitken, and B.W.P. Wells, "Chronic *Toxoplasma* Infections and Familiarity-Novelty Discrimination in the Mouse," *Annals of Tropical Medicine and Parasitology* 74 (1980): 145–50.

In addition, Hutchison: J. Hay et al., "The Effect of Congenital and Adult-Acquired *Toxoplasma* Infections on Activity and Responsiveness to Novel Stimulation in Mice," *Annals of Tropical Medicine and Parasitology* 77 (1983): 483–95.

More ominously, Otto: V. O. Jírovec, "Die Toxoplasmose-Forschung in der Tscheehoslowakei," *Tropenmedizin Und Parasitologie* 7, no. 3 (September 1956): 281–82.

"The parasite can't": Flegr interview.

Indeed, in France: K. McAuliffe, "How Your Cat Is Making You Crazy," *Atlantic*, March 2013.

62   *To test the manipulation hypothesis:* Flegr interview.

But people with the latent infection: J. Flegr and I. Hrdý, "Influence of Chronic Toxoplasmosis on Some Human Personality Factors," *Folia Parasitology* 41 (1994): 122–26.

Skeptical of his: J. Flegr et al., "Induction of Changes in Human Behaviour by the Parasitic Protozoan *Toxoplasma gondii*," *Parasitology* 113 (1996): 49–54.

Participants in one: J. Lindova et al., "Gender Differences in Behavioural Changes Induced by Latent Toxoplasmosis," *International Journal for Parasitology* 36 (2006): 1485–92.

63   *"They often wore":* Flegr interview.

On a computerized: J. Havlicek et al., "Decrease of Psychomotor Performance in Subjects with Latent 'Asymptomatic' Toxoplasmosis," *Parasitology* 122 (2001): 515.

Flegr launched a: J. Flegr et al., "Increased Risk of Traffic Accidents in Subjects with Latent Toxoplasmosis: A Retrospective Case-Control Study," *BioMed Central Infectious Diseases* 2 (July 2002): 11.

Since the strength: J. Flegr et al., "Increased Incidence of Traffic Accidents in *Toxoplasma*-Infected Military Drivers and Protective Effect RhD Molecule Revealed by a Large-Scale Prospective Cohort Study," *BioMed Central Infectious Diseases* 9 (May 2009): 72.

"I estimate that": Flegr interview.

66   *When we were:* J. Horáček et al., "Latent Toxoplasmosis Reduces Gray Matter Density in Schizophrenia but Not in Controls: Voxel-Based-Morphometry (Vbm) Study," *World Journal of Biological Psychiatry* 13 (2012): 501.

Jiří Horáček: J. Horáček, interview by the author, September 21, 2011.

His research showing: M. Aslan et al., "Higher Prevalence of Toxoplasmosis in Victims of Traffic Accidents Suggest Increased Risk of Traffic Accident in *Toxoplasma*-Infected Inhabitants of Istanbul and Its Suburbs," *Forensic Science International* 187, nos. 1–3 (May 30, 2009): 103. Also see K. Yereli, I. C. Balcioglu, and A. Ozbilgin, "Is *Toxoplasma gondii* a Potential Risk for Traffic Accidents in Turkey?," *Forensic Science International* 163 (2006): 34, and M. L. Galván-Ramírez, L. V. Sánchez-Orozco, and L. Rocío Rodríguez, "Seroepidemiology of *Toxoplasma gondii* Infection in Drivers Involved in Road Traffic Accidents in the Metropolitan Area of Guadalajara, Jalisco, Mexico," *Parasites and Vectors* 6 (2013): 294.

Still another study: C. Alvarado-Esquivel et al., "High Seroprevalence of *Toxoplasma gondii* Infection in a Subset of Mexican Patients with Work Accidents and Low Socioeconomic Status," *Parasites and Vectors* 5 (2012): 13.

67 *yet Flegr himself:* Jaroslav Flegr, interview by the author, Massa Marittima, Italy, March 20, 2012.

And when he: J. Lindová, L. Příplatová, and J. Flegr, "Higher Extraversion and Lower Conscientiousness in Humans Infected with Toxoplasma," *European Journal of Personality* 26 (2012): 285.

68 *Webster would not:* Webster interview.

By the time: Joanne Webster, interview by the author, May 2012.

Then, studying rats: Webster interview, summer 2011. Also see M. Berdoy, J. P. Webster, and D. W. Macdonald, "Fatal Attraction in *Toxoplasma*-Infected Rats: A Case of Parasite Manipulation of Its Mammalian Host," *Proceedings of the Royal Society B* 267 (2000): 1591–94.

69 *For most of the:* Glenn McConkey, interview by the author, September 16, 2011, and May 1, 2012.

70 *The gene, he was:* E. Gaskell et al., "A Unique Dual Activity Amino Acid Hydroxylase in *Toxoplasma gondii*," *PLoS One* 4, no. 3 (March 2009): e4801.

By 2011, they: E. Prandovszky et al., "The Neurotropic Parasite *Toxoplasma gondii* Increases Dopamine Metabolism," *PLoS One* 6, no. 9 (September 2011): e23866.

For example, previous: Webster interview, summer 2011.

Meanwhile, a team: Robert Sapolsky, interview by the author, September 13, 2011.

71 *"This is flabbergasting":* R. Sapolsky, "Bugs in the Brain," *Scientific American* (March 2003).

At a salon-style forum: R. Sapolsky, "Toxo: A Conversation with Robert Sapolsky," *Edge,* December 4, 2009, http://edge.org/conversation/robert_sapolsky-toxo.

The British reports: Ibid.

without having secured: Patrick House, interview by the author, Palo Alto, California, July 18, 2014.

His team set out: Sapolsky interview, summer 2011 and September 13, 2011.

72   T. gondii *travels:* Ajai Vyas, interview by the author, summer 2011.

In the middle of: Flegr interview, summer 2011 and October 2011. Also see J. Flegr et al., "Fatal Attraction Phenomenon in Humans — Cat Odour Attractiveness Increased for *Toxoplasma*-Infected Men," *PLoS* 5, no. 11 (November 2011): e1389.

73   *Soon Vyas and I:* Vyas interview.

For example, male: Katty Kay and Claire Shipman, "The Confidence Gap," *Atlantic,* May 2014.

When the excess: S. A. Hari Dass and A. Vyas, "*Toxoplasma gondii* Infection Reduces Predator Aversion in Rats Through Epigenetic Modulation in the Host Medial Amygdala," *Molecular Ecology* 23, no. 4 (December 2014): 6114–22, doi: 10.1111/mec.12888.

No less wondrous: Doruk Golcu, Rahiwa Z. Gebre, and Robert M. Sapolsky, "*Toxoplasma gondii* Influences Aversive Behaviors of Female Rats in an Estrus Cycle Dependent Manner," *Physiology and Behavior* 135 (2014): 98–103.

74   *But toxoplasma never ceases:* Doruk Golcu, interview by the author, November 10, 2015.

In 2013, Sapolsky retired: House interview.

He reports that: Andrew Evans, interview by the author, May 13 and May 15, 2013, and March 24 and March 25, 2014.

75   *Sapolsky also urges:* Sapolsky interview, summer 2011 and September 13, 2011.

Webster takes a: Webster interview, summer 2011.

76   *House explained en route:* House interview.

78   *"This has not been":* E. Fuller Torrey, interview by the author, Bethesda, Maryland, January 22, 2013.

79   *"Textbooks today still":* E. Fuller Torrey, interview by the author, July 28, 2011.

a conclusion he: E. Fuller Torrey and Judy Miller, *The Invisible Plague: The Rise of Mental Illness from 1750 to the Present* (New Brunswick, NJ: Rutgers University Press, 2001).

The disease "is so striking": Torrey interview, July 28, 2011.

Most compelling, people: E. F. Torrey, J. J. Bartko, and R. H. Yolken, "*Toxoplasma gondii* and Other Risk Factors for Schizophrenia: An Update," *Schizophrenia Bulletin* 38, no. 3 (2012): 642–47, doi:10.1093/schbul/sbs.

Human genome findings: Torrey interview, July 28, 2011; Robert Yolken, interview by the author, July 25, 2011.

80   *Adding further fuel:* Teodor Postolache, interview by the author, Baltimore, Maryland, January 17, 2013.

Across twenty-five nations: V. J. Ling et al., "*Toxoplasma gondii* Seropositivity and Suicide Rates in Women," *Journal of Nervous and Mental Disease* 199, no. 7 (July 2011).

In concert with: M. G. Pedersen et al., "*Toxoplasma gondii* Infection and Self-Directed Violence in Mothers," *Archives of General Psychiatry* 69, no. 11 (November 2012): 1124–29.

Turkey: F. Yagmur et al., "May *Toxoplasma gondii* Increase Suicide Attempt? Preliminary Results in Turkish Subjects," *Forensic Science International* 199, nos. 1–3 (June 15, 2010): 15–17, doi: 10.1016/j.forsciint.2010.02.020.

Sweden: Y. Zhang et al., "*Toxoplasma gondii* Immunoglobulin G Antibodies and Nonfatal Suicidal Self-Directed Violence," *Journal of Clinical Psychiatry* 73, no. 8 (2012): 1069–76, doi: 10.4088/JCP.11m07532.

Baltimore/Washington area: T. Arling, R. H. Yolken, and M. Lapidus, "*Toxoplasma gondii* Antibody Titers and History of Suicide Attempts in Patients with Recurrent Mood Disorders," *Journal of Nervous and Mental Disease* 197, no. 2 (December 2009): 905.

81 *The reason is:* T. B. Cook et al., "'Latent' Infection with *Toxoplasma gondii*: Association with Trait Aggression and Impulsivity in Healthy Adults," *Journal of Psychiatric Research* 60 (January 2015): 87–94.

82 *Postolache cherishes:* Postolache interview.

5. DANGEROUS LIAISONS

83 *The idea for the:* C. Reiber, interview by the author, August 18, 2011, and January 13, 2013.

84 *Then Moore got an inspiration:* Ibid.; and Janice Moore, interview by the author, fall 2011 and January 6, 2015.

85 *The researchers tracked:* C. Reiber et al., "Changes in Human Social Behavior in Response to a Common Vaccine," *Annals of Epidemiology* 20, no. 10 (October 2010), doi: 10.1016/j.annepidem.2010.06.014.

"People who had": C. Reiber, interview by the author, January 15, 2013; Moore interview, fall 2011 and January 6, 2015.

On a blog sponsored: Kristi McGuire, "Traffic: Carl Zimmer and W. Ian Lipkin," *The Chicago Blog,* April 11, 2015, http://pressblog.uchicago.edu/2011/05/03 /traffic-carl-zimmer-and-w-ian-lipkin.html.

In a more recent conversation: Reiber interview, January 15, 2013.

86 *At the University of Montpellier:* Frédéric Thomas, interview by the author, Massa Marittima, Italy, March 19, 2012.

A sharp rise in: Charles Rupprecht, interview by the author, December 12, 2012; and see B. Wasik and M. Murphy, *Rabid: A Cultural History of the World's Most Diabolical Virus* (New York: Penguin, 2012), 9; J. K. Dutta, "Excessive Libido in a Woman with Rabies," *Postgraduate Medical Journal* 72 (1996): 554; and A. M. Gardner, "An Unusual Case of Rabies," *Lancet* 296, no. 7671 (1970): 523.

In earlier centuries: K. Kete, *The Beast in the Boudoir: Petkeeping in Nineteenth-Century Paris* (Berkeley: University of California Press, 1994), 101–2.

Men may experience: Wasik and Murphy, *Rabid,* 9.

Of course, rabid: Rupprecht interview.

87   *Although those blessed:* "Rabies," World Health Organization, updated September 2015, http://www.who.int/mediacentre/factsheets/fs099/en/.

develop the first vaccine: Wasik and Murphy, *Rabid,* 10.

Instead, it creeps along: Ibid., 8.

89   *In a case reported in India:* S. Senthilkumaran et al., "Hypersexuality in a 28-Year-Old Woman with Rabies," *Archives of Sexual Behavior* 40, no. 6 (2011): 1327–28.

91   *in their underground vaults:* J. Gómez-Alonso, "Rabies: A Possible Explanation for the Vampire Legend," *Neurology* 51 (1998): 856–59.

At least the canis: Celia Holland, interview by the author, November 13, 2012.

93   *In an early study:* M.R.H. Taylor et al., "The Expanded Spectrum of Toxocaral Disease," *Lancet* (March 26, 1988): 692.

A study published: M. G. Walsh and M. A. Haseeb, "Reduced Cognitive Function in Children with Toxocariasis in a Nationally Representative Sample of the United States," *International Journal for Parasitology* 42 (2012): 1159–63, http://www.ncbi.nlm.nih.gov/pubmed/23123274.

94   *Like Holland, Walsh was:* Michael Walsh, interview by the author, November 13, 2012.

95   *For example, Holland:* Holland interview.

97   *One strategy:* Walsh interview.

98   *More than fourteen hundred parasites:* Charles Nunn, interview by the author, April 15, 2015.

6. GUT FEELINGS

99   *The first major census:* G. Kolata, "In Good Health? Thank Your 100 Trillion Bacteria," *New York Times,* June 13, 2012, http://www.nytimes.com/2012/06/14/health/human-microbiome-project-decodes-our-100-trillion-good-bacteria.html.

100   *The amount of genetic:* J. F. Cryan and T. G. Dinan, "Mind-Altering Microorganisms: The Impact of the Gut Microbiota on Brain and Behavior," *Nature Reviews Neuroscience* 13, no. 10 (2012): 702, doi: 10.1038 /nrn3346.

Some of these microbes: D. Grady, "Study Sees Bigger Role for Placenta in Newborns' Health," *New York Times,* May 21, 2014, http://www.nytimes.com/2014/05/22/health/study-sees-bigger-role-for-placenta-in-newborns-health.html.

But the biggest wave: Stephen Collins, interview by the author, January 7, 2013.

Children and adults typically: C. Lozupone et al., "Diversity, Stability and Resilience of the Human Gut Microbiota," *Nature* 489 (September 13, 2012): 220–23, doi: 10.1038/nature11550.

101   *The same species of bacteria:* M. J. Blaser, interview by the author, December 18, 2012.

In the gut: Kolata, "In Good Health."

They also churn: Cryan and Dinan, "Mind-Altering Microorganisms," 704.

To varying degrees: Ibid., 701–9; and P. Forsythe et al., "Mood and Gut Feelings," *Brain, Behavior, and Immunity* 24 (2010): 9–16, doi:10.1016/j.bbi.2009.05.058.

Some psychoactive compounds made: A. Hadhazy, "Think Twice: How the Gut's 'Second Brain' Influences Mood and Well-Being," *Scientific American* (February 12, 2010), http://www.scientificamerican.com/article/gut-second-brain/.

In the process: Lindsay Borthwick, "Microbiome and Neuroscience: The Mind-Bending Power of Bacteria," Kavli Foundation, Winter 2014, http://www.kavlifoundation.org/science-spotlights/mind-bending-power-bacteria.

Perhaps related to: M. Almond, "Depression and Inflammation: Examining the Link," *Current Psychiatry* 12, no. 6 (June 2013): 24–32. Also see A. Naseribafrouei et al., "Correlation Between the Human Fecal Microbiota and Depression," *Neurogastroenterology and Motility* 26 (2014): 1155–62.

102   *Indeed, 50 to 80 percent:* Collins interview.

More surprising, specific: M. Wenner Moyer, "Gut Bacteria May Play a Role in Autism," *Scientific American* (August 14, 2014).

Rodents that display: J. Gilbert et al., "Toward Effective Probiotics for Autism and Other Neurodevelopmental Disorders," *Cell* 155, no. 7 (2013): 1446, http://dx.doi.org/10.1016/j.cell.2013.11.035.

The most striking proof: S. Collins, M. Surette, and P. Bercik, "The Interplay Between the Intestinal Microbiota and the Brain," *Nature Reviews Microbiology* 10, no. 11 (2012): 735–42, doi: 10.1038/nrmicro2876.

103   *In addition, germfree mice:* M. G. Gareau et al., "Bacterial Infection Causes Stress-Induced Memory Dysfunction in Mice," *Gut* 60, no. 3 (2011): 307–17, doi: 10.1136/gut.2009.202515.

Just as in the case: R. Heijtz et al., "Normal Gut Microbiota Modulates Brain Development and Behavior," *Proceedings of the National Academy of Sciences of the United States of America* 108, no. 7 (2011): 3047–52, doi: 10.1073/pnas.1010529108.

The same team that: Gareau et al., "Bacterial Infection Causes Stress-Induced Memory Dysfunction," 307.

104   *This is the domain:* J. Cryan, interview by the author, December 10, 2012. Also see J. Bravo et al., "Ingestion of *Lactobacillus* Strain Regulates Emotional Behavior and Central GABA Receptor Expression in a Mouse Via the Vagus

Nerve," *Proceedings of the National Academy of Sciences of the United States of America* 108, no. 38 (2011): 16050–55, doi: 10.1073/pnas.1102999108.

106   *Cryan's team and other:* F. Dickerson, interview by the author, March 26, 2014.
symptoms of functional gastrointestinal disorders: Yoshihisa Urita et al., "Continuous Consumption of Fermented Milk Containing *Bifidobacterium bifidum* YIT 10347 Improves Gastrointestinal and Psychological Symptoms in Patients with Functional Gastrointestinal Disorders," *Bioscience of Microbiota, Food and Health* 34, no. 2 (2015): 37–44. Also C. Janssen, interview by the author, June 25, 2013, and December 7, 2015.
A sprinkling of clinical: F. Indrio et al., "Prophylactic Use of a Probiotic in the Prevention of Colic, Regurgitation, and Functional Constipation: A Randomized Clinical Trial," *JAMA Pediatrics* 168, no. 3 (2014): 228–33, doi: 10.1001/jamapediatrics.2013.4367. Also see B. Chumpitazi and R. J. Shulman, "Five Probiotic Drops a Day to Keep Infantile Colic Away?," *JAMA Pediatrics* 168, no. 3 (2014): 204–5, doi:10.1001/jamapediatrics.2013.5002.
In one trial: F. Savino et al., "*Lactobacillus reuteri* (American Type Culture Collection Strain 55730) Versus Simethicone in the Treatment of Infantile Colic: A Prospective Randomized Study," *Pediatrics* 119, no. 1 (2007): e124–e130.
For example, a randomized: M. Messaoudi et al., "Assessment of Psychotropic-like Properties of a Probiotic Formulation (*Lactobacillus helveticus* R0052 and *Bifidobacterium longum* R0175) in Rats and Human Subjects," *British Journal of Nutrition* 105, no. 5 (2011): 755–64, doi: 10.1017/S0007114510004319.

107   *Conducted at UCLA:* K. Tillisch et al., "Consumption of Fermented Milk Product with Probiotic Modulates Brain Activity," *Gastroenterology* 144, no. 7 (2013): 1394–1401, doi: 10.1053/j.gastro.2013.02.043.
"We have to be": E. Mayer, interview by the author, September 13, 2013, and April 15, 2014.

109   *Presently, the probiotics:* Cryan interview.
That process will be: Mayer interview.

7. MY MICROBES MADE ME FAT

111   *Behold two mice:* See R. Marantz Henig, "Fat Factors," *New York Times*, August 13, 2006, and F. Bäckhed et al., "The Gut Microbiota as an Environmental Factor That Regulates Fat Storage," *Proceedings of the National Academy of Science* 101, no. 44 (November 2, 2004): 15718–23.
Intestinal bacteria regulate: M. J. Blaser, interview by the author, December 18, 2012. Also see M. J. Blaser, "Stop the Killing of Beneficial Bacteria," *Nature* 476 (August 25, 2011): 293–94, and P. L. Jeffrey et al., "Endocrine Impact of *Helicobacter pylori:* Focus on Ghrelin and Ghrelin O-Acyltransferase," *World Journal*

of Gastroenterology 17, no. 10 (March 14, 2011): 1249–60, doi: 10.3748/wjg.v17 .i10.1249.

It's also suspected: John F. Cryan, interview by the author, December 10, 2012. Also see J. M. Kinross et al., "The Human Gut Microbiome: Implications for Future Health Care," *Current Gastroenterology Reports* 10 (2008): 396–403.

112 *In 2006, Gordon's team:* P. J. Turnbaugh et al., "An Obesity-Associated Gut Microbiome with Increased Capacity for Energy Harvest," *Nature* 444, no. 7122 (2006): 1027–31, doi:10.1038/nature05414.

Obese and thin humans: R. E. Ley et al., "Microbial Ecology: Human Gut Microbes Associated with Obesity," *Nature* 444, no. 7122 (2006): 1022–23, doi: 10.1038/4441022a.

To untangle cause: V. K. Ridaura et al., "Gut Microbiota from Twins Discordant for Obesity Modulate Metabolism in Mice," *Science* 341, no. 6150 (2013), doi: 10.1126/science.1241214.

113 *More recently, Gordon's:* C. Wallis, "Gut Reactions," *Scientific American* 310, no. 6 (June 2014): 30–33.

But paradoxically, their: A. W. Walker and J. Parkhill, "Fighting Obesity with Bacteria," *Science* 341, no. 1069 (2013), doi: 10.1126/science.1243787.

With the goal of: C. Ostrom, "Wonder Cure for Gut: FDA Allows Fecal Transplants," *Seattle Times,* October 26, 2013.

114 *The move, however, was:* Wallis, "Gut Reactions."

donors are rigorously screened: Stephen Collins, interview by the author, January 7, 2013.

Noting that gut bacteria: Cryan interview.

Collins — who did the: Collins interview.

Meanwhile, Gordon and: G. Kolata, "Gut Bacteria from Thin Humans Can Slim Mice Down," *New York Times,* September 5, 2013.

Not just anal but: See J. R. Cryan and T. G. Dinan, "Mind-Altering Microorganisms: The Impact of the Gut Microbiota on Brain and Behavior," *Nature Reviews Neuroscience* 13 (October 2012): 701–12, and C. Lozupone et al., "Diversity, Stability and Resilience of the Human Gut Microbiota," *Nature* 489 (September 13, 2012): 221.

115 *As Gordon told:* Wallis, "Gut Reactions."

Antibiotics, by depleting: Blaser, "Stop the Killing."

Lagging far behind: S. Tavernise, "F.D.A. Restricts Antibiotics Use for Livestock," *New York Times,* December 11, 2013.

Another farmer's: Blaser, "Stop the Killing."

To answer that: Wallis, "Gut Reactions."

116 *A decade-long study:* Bloomberg School of Public Health, Johns Hopkins University, "Children Who Take Antibiotics Gain Weight Faster Than Kids Who Don't," news release, October 21, 2015.

Ironically, a celebrated: Blaser, "Stop the Killing."

But there may: Kate Murphy, "In Some Cases, Even Bad Bacteria May Be Good," *New York Times,* October 31, 2011.

117   *Babies delivered by:* Wallis, "Gut Reactions."

Better focusing the: Murphy, "In Some Cases, Even Bad Bacteria May Be Good."

118   *Experts also warned:* Emeran Mayer, interview by the author, September 13, 2013; and Faith Dickerson, interview by the author, March 26, 2014.

In one of the largest: D. Mozaffarian et al., "Changes in Diet and Lifestyle and Long-Term Weight Gain in Women and Men," *New England Journal of Medicine* 364 (June 23, 2011): 2392–404.

119   *The lead investigator, Frank:* J. E. Brody, "Still Counting Calories? Your Weight-Loss Plan May Be Outdated," *New York Times,* July 18, 2011.

But when bacteria: M. J. Blaser, "Who Are We? Indigenous Microbes and the Ecology of Human Diseases," *European Molecular Biology Organization Reports* 7, no. 10 (2006): 957.

According to microbiome: Mark Lyte, interview by the author, March 19, 2014.

120   *Which is why without:* Collins interview.

"These microbes have": Mayer interview.

8. HEALING INSTINCT

124   *To turn up the body's thermostat:* G. Pacheco-López and F. Bermúdez-Rattoni, "Brain-Immune Interactions and the Neural Basis of Disease-Avoidant Ingestive Behavior," *Philosophical Transactions of the Royal Society B* 366 (2011): 3397.

Fever is so important: M. J. Perrot-Minnot and F. Cézilly, "Parasites and Behaviour," in *Ecology and Evolution of Parasitism,* ed. F. Thomas, J. F. Guégan, and F. Renaud (Oxford: Oxford University Press, 2009), 61; also see Randolph M. Nesse and George C. Williams, *Why We Get Sick: The New Science of Darwinian Medicine* (New York: Vintage, 1994), 27.

Lest anyone doubt: R. H. McCusker, *Journal of Experimental Biology* Conference on Neural Parasitology, Tuscany, Italy, March 19, 2012; R. H. McCusker and K. W. Kelley, "Immune-Neural Connections: How the Immune System's Response to Infectious Agents Influences Behavior," *Journal of Experimental Biology* 216 (2013): 84–98, doi: 10.1242/jeb.073411.

Sometimes fever: Nesse and Williams, *Why We Get Sick,* 37. Giulia Enders, *Gut: The Inside Story of Our Body's Most Underrated Organ* (Vancouver, Berkeley: Greystone Books, 2015), part 2; subheading: vomiting.

125   *Such copycat behavior:* Rachel Herz, *That's Disgusting: Unraveling the Mysteries of Repulsion* (New York: W. W. Norton, 2012), 73.

saw them up close: Benjamin Hart, interview by the author, Davis, California, September 6, 2013.

126 *humans and birds alone:* Cindy Engel, *Wild Health: Lessons in Natural Wellness from the Animal Kingdom* (Boston: Houghton Mifflin, 2002), 109.
Tabanid flies: B. L. Hart, "Behavioral Adaptations to Pathogens and Parasites: Five Strategies," *Neuroscience and Biobehavioral Reviews* 14, no. 3 (1990): 276.
Rodents spend one-third: Ibid., 277.
Blue herons peck: Engel, *Wild Health,* 111.
Horses and other hoofed: Hart, "Behavioral Adaptations to Pathogens and Parasites," 277–79.

127 *Those herons:* Engel, *Wild Health,* 111.
And in a study the Harts conducted: B. Hart, "Behavioural Defences in Animals Against Pathogens and Parasites: Parallels with the Pillars of Medicine in Humans," *Philosophical Transactions of the Royal Society B* 366 (December 2011): 3407.
Pest-avoidance strategies: Benjamin and Lynette Hart, interview by the author, Davis, California, September 9, 2013.

128 *Wood mice and long-tailed macaques:* Engel, *Wild Health,* 113; and Valerie Curtis, *Don't Look, Don't Touch, Don't Eat: The Science Behind Revulsion* (Chicago: University of Chicago Press, 2013), Kindle edition, chapter 2.
Other animals outsource: Hart, "Behavioral Adaptations to Pathogens and Parasites," 279.
One of the bird's specialties: Benjamin Hart interview.
One of their many tricks: Hart, "Behavioural Defences in Animals," 3408.

129 *the removal of rodents' salivary glands:* L. Bodner et al., "The Effect of Selective Desalivation on Wound Healing in Mice," *Experimental Gerontology* 26, no. 4 (1991): 383–86.
a sheet of human cells grown in culture: Federation of American Societies for Experimental Biology press release, "Licking Your Wounds: Scientists Isolate Compound in Human Saliva That Speeds Wound Healing," *Science Digest,* July 24, 2008, http://www.sciencedaily.com/releases/2008/07/0807230948 41.htm.
Like primates today: Benjamin and Lynette Hart interview.
Contact with fecal: Curtis, *Don't Look, Don't Touch, Don't Eat,* chapter 1.

130 *Poop poses just as broad:* Engel, *Wild Health,* 78–79.
Mole rats and: Perrot-Minnot and Cézilly, "Parasites and Behaviour," 53.
Cows, sheep, and: Hart, "Behavioral Adaptations to Pathogens and Parasites," 277.
Wolves, hyenas, and: Benjamin Hart interview.
Fish, too, have: Curtis, *Don't Look, Don't Touch, Don't Eat,* chapter 2.

131 *bee poop for chemical warfare:* Natalie Angier, "Nature's Waste Management Crews," *New York Times,* May 25, 2015.
Bullfrog tadpoles can: Perrot-Minnot and Cézilly, "Parasites and Behaviour," 54.
Killifish treated with: Curtis, *Don't Look, Don't Touch, Don't Eat,* chapter 2.

Apes show similar: C. Zimmer, "Is Patriotism a Subconscious Way for Humans to Avoid Disease?," *New York Times,* February 18, 2009.

When an ant gets a deadly fungus: Sindya N. Bhanoo, "Tending a Sick Comrade Has Benefits for Ants," *New York Times,* April 9, 2012.

132 *Mothers of newly weaned:* Benjamin Hart interview.

Indeed, claylike compounds: Tai Viinikka, "About Kids Health: The Hazards and Benefits of Eating Dirt," Hospital for Sick Children, Toronto, Ontario, Canada, May 16, 2013, http://www.aboutkidshealth.ca/en/news/newsand features/pages/the-hazards-and-benefits-of-eating-dirt.aspx.

Some species of monkeys: Benjamin Hart interview.

133 *Complementing that view:* Viinikka, "About Kids Health."

134 *Like men who pump iron:* Hart, "Behavioral Adaptation to Pathogens and Parasites," 288.

Female mice are: Curtis, *Don't Look, Don't Touch, Don't Eat,* chapter 2.

As part of their mating ritual: Hart, "Behavioral Adaptations to Pathogens and Parasites," 287.

135 *With that in mind:* S. W. Gangestad and D. M. Buss, "Pathogen Prevalence and Human Mate Preferences," *Ethology and Sociobiology* 14 (1993): 89–96.

In a British study: A. C. Little, B. C. Jones, and L. M. DeBruine, "Exposure to Visual Cues of Pathogen Contagion Changes Preferences for Masculinity and Symmetry in Opposite-Sex Faces," *Proceedings of the Royal Society B* 278 (2011): 813–14, doi: 10.1098/rspb.2010.1925.

which is still contentious: "HLA Gene Family," Genetics Home Reference, accessed March 11, 2015, http://ghr.nlm.nih.gov/geneFamily=hla.

In the 1990s: Daniel M. Davis, *The Compatibility Gene: How Our Bodies Fight Disease, Attract Others, and Define Our Selves* (New York: Oxford University Press, 2014), 137–40.

136 *end of a family lineage:* J. M. Tyburg and S. W. Gangestad, "Mate Preferences and Infectious Disease: Theoretical Considerations and Evidence in Humans," *Philosophical Transactions of the Royal Society B* 366 (2011): 3383, doi: 10.1098/rstb.2011.0136.

That liability is greatly: Hart, "Behavioral Adaptations to Pathogens and Parasites," 288.

Inspired by the animal research: C. Wedekind et al., "MHC-Dependent Mate Preferences in Humans," *Proceedings of the Royal Society B* 260 (1995): 245–49.

More recently, Wedekind: M. Milinsky and C. Wedekind, "Evidence for MHC-Correlated Perfume Preferences in Humans," *Behavioural Ecology* 12 (2001): 140–49.

137 *Meshing nicely with:* C. Ober et al., "HLA and Mate Choice in Humans," *American Journal of Human Genetics* 61 (1997): 497–504; and R. Chaix, C. Cao, and P. Donnelly, "Is Mate Choice in Humans MHC-Dependent?," *PLoS Genetics* 4, no. 9 (September 12, 2008): e1000184, doi: 10.1371/journal.pgen.1000184.

In a related line of research: F. Prugnolle et al., "Pathogen-Driven Selection and Worldwide HLA Class I Diversity," *Current Biology* 15 (2005): 1022–27.

Also complementing Wedekind's work: Herz, *That's Disgusting,* 170; and R. S. Herz and M. Inzlicht, "Sex Differences in Response to Physical and Social Factors Involved in Human Mate Selection," *Evolution and Human Behavior* 23 (2002): 359–64.

Studies of the Yoruba: Chaix, Cao, and Donnelly, "Is Mate Choice in Humans MHC-Dependent?"

Strong support for this view: Herz, *That's Disgusting,* 171.

138 *Eons ago:* Matt Ridley, "The Advantage of Sex," *New Scientist,* December 4, 1993.

After briefly falling: M. Scudellari, "The Sex Paradox," *Scientist,* July 1, 2014.

What makes sex: Ridley, "The Advantage of Sex."

If you're human "HLA Gene Family," Genetics Home Reference.

Even as Ebola: Joel Achenbach, Lena H. Sun, and Brady Dennis, "The Ominous Math of the Ebola Epidemic," *Washington Post,* October 9, 2014, http://www .washingtonpost.com/national/health-science/the-ominous-math-of-the -ebola-epidemic/2014/10/09/3cad9e76-4fb2-11e4-8c24-487e92bc997b_story .html.

139 *Many of these lucky:* R. Rettner, "How Do People Survive Ebola?," *Live Science,* August 5, 2014, http://www.livescience.com/47203-ebola-how-people-survive .html.

Precisely because we aren't clones: Ridley, "The Advantage of Sex."

Just think: R. M. Nesse and G. C. Williams, "Evolution and the Origins of Disease," *Scientific American* (November 1998). Also see "Teach Evolution and Make It Relevant," University of Montana, http://evoled.dbs.umt.edu/lessons /background.htm.

Sleep is a puzzle: F. Thomas et al., "Can We Understand Modern Humans Without Considering Pathogens?," *Evolutionary Applications* 5, no. 4 (June 2012): 373.

140 *A novel theory posits:* B. T. Preston et al., "Parasite Resistance and the Adaptive Significance of Sleep," *BMC Evolutionary Biology* 9, no. 7 (January 9, 2009): 1–9, doi: 10.1186/1471-2148-9-7.

141 *Intriguingly, reports Benjamin Hart:* B. L. Hart, "The Evolution of Herbal Medicine: Behavioural Perspectives," *Animal Behaviour* 70 (2005): 983, doi: 10.1016/j.anbehav.2005.03.005.

Some species of tiger moth: E. A. Bernays and M. S. Singer, "Insect Defenses: Taste Alteration and Endoparasites," *Nature* 436 (July 28, 2005): 476.

142 *Healthy bystanders:* M. A. Huffman, "Current Evidence for Self-Medication in Primates: A Multidisciplinary Perspective," *Yearbook of Physical Anthropology* 40 (1997): 178.

Goodall recorded: Hart, "The Evolution of Herbal Medicine," 983.

Herbivores are: Engel, *Wild Health,* 84.

People in some villages: Cheryl L. Dybas, "*Aframomum melegueta:* Gorilla Staple Adds Spice to New Drugs," *Washington Post,* November 27, 2006.

In Uganda: Huffman, "Current Evidence for Self-Medication in Primates," 173.

Many of these seasonings: P. W. Sherman and J. Billing, "Darwinian Gastronomy: Why We Use Spices," *BioScience* 49 (1999): 455.

143 *Oddly, they found:* Ibid., 458.

Since food, especially meat: Ibid., 455.

A subsequent study: P. W. Sherman and G. A. Hash, "Why Vegetable Dishes Are Not Very Spicy," *Evolution and Human Behavior* 22, no. 3 (May 2001): 147–63.

144 *Indeed, garlic, turmeric:* Hart, "The Evolution of Herbal Medicine," 977–79.

In the last trimester: Annie Murphy Paul, *Origins: How the Nine Months Before Birth Shape the Rest of Our Lives* (New York: Free Press, 2010), 22.

A passion for spices: Sherman and Billing, "Darwinian Gastronomy," 461–62.

145 *Indeed, the harnessing of fire:* Rachael Moeller Gorman, "Cooking Up Bigger Brains," *Scientific American* (December 16, 2007), http://www.scientific american.com/article/cooking-up-bigger-brains/.

As Benjamin Hart explains: Hart, "Behavioural Defences in Animals," 3409.

The consumption of leaves: M. A. Huffman and J. M. Caton, "Self-Induced Increase of Gut Motility and the Control of Parasitic Infections in Wild Chimpanzees," *International Journal of Primatology* 22, no. 3 (2001): 329–46. Also M. A. Huffman, interview by the author, December 1, 2015.

Some birds and rodents: Hart, "Behavioral Adaptations to Pathogens and Parasites," 280–81.

146 *One plant favored:* Engel, *Wild Health,* 123.

The dusky-footed wood rat: Perrot-Minnot and Cézilly, "Parasites and Behaviour," 57.

Entomologists use bay: John Smart, "Number 4A: Insects," *British Museum of Natural History Instructions for Collectors* (London: Trustees of the British Museum, 1963).

In field experiments: Hart, "Behavioral Adaptations to Pathogens and Parasites," 281.

Nature's pharmacopoeia: Huffman, "Current Evidence for Self-Medication in Primates," 190.

In Panama: Engel, *Wild Health,* 115–18.

147 *It shields the gut:* "Dr. Sera Young, Cornell University — the Urge to Eat Dirt," *Academic Minute,* WAMC Northeast Public Radio, December 4, 2012, http://wamc.org/post/dr-sera-young-cornell-university-urge-eat-dirt. Also see Sera L. Young, *Craving Earth: Understanding Pica* (New York: Columbia University Press, 2012), Kindle edition, chapter 9, and Engel, *Wild Health,* 64–70.

Both humans and: Young, *Craving Earth,* chapter 9.

The earth craved: Susan Allport, "Women Who Eat Dirt," *Gastronomica* (Spring 2002): 17.

Owing to the molecular: Young, *Craving Earth,* chapter 9; Engel, *Wild Health,* 62–70.

Extensive field observations: Engel, *Wild Health,* 63–70.

148   *People have been:* Young, *Craving Earth,* chapter 3.

The practice, called: Sera L. Young, interview by the author, November 23, 2015.

Many geophagists report: Marc Lallanilla, "Eating Dirt: It Might Be Good for You," ABC News, October 3, 2005, http://abcnews.go.com/Health/Diet /story?id=1167623&page=1.

"When I'm pregnant": Young, *Craving Earth,* chapter 1.

Highlighting clay's importance: Ibid., chapter 9.

By far the largest: Ibid., chapter 1.

This makes sense: Ibid., chapter 9.

Not surprisingly, women: Ibid., chapter 1.

149   *During the first trimester:* Thomas et al., "Can We Understand Modern Humans Without Considering Pathogens?," 374–75; Meredith F. Small, "The Biology of Morning Sickness," *Discover,* September 1, 2000, http://discover magazine.com/2000/sep/featbiology.

9. THE FORGOTTEN EMOTION

151   *When Valerie Curtis:* Valerie Curtis, interview by the author, July 1, 2013.

152   *Curtis doesn't know:* Valerie Curtis, *Don't Look, Don't Touch, Don't Eat: The Science Behind Revulsion* (Chicago: University of Chicago Press, 2013), Kindle edition, chapter 2.

As we learn: Curtis interview.

153   *She and her students:* Curtis, *Don't Look, Don't Touch, Don't Eat,* chapter 1.

I could go on and on: Curtis interview.

154   *In centuries past:* Charles Darwin, *The Expression of the Emotions in Man and Animals* (London: Penguin Classics, 1872), Kindle edition, chapter 11.

155   *Interestingly in that regard:* Curtis, *Don't Look, Don't Touch, Don't Eat,* chapter 1.

Despite Darwin's perspicacity: Darwin, *The Expression of the Emotions,* chapter 11.

More than a century: Curtis, *Don't Look, Don't Touch, Don't Eat,* chapter 1.

He theorized: Paul Rozin, interview by the author, Philadelphia, January 21, 2013.

Among my favorites: M. Oaten, R. J. Stevenson, and T. I. Case, "Disgust as a Disease-Avoidance Mechanism," *Psychological Bulletin* 135, no. 2 (2009): 312.

156   *Rozin's pioneering investigations:* Rozin interview. For good overview articles

on disgust, see Oaten, Stevenson, and Case, "Disgust as a Disease-Avoidance Mechanism," 303–21, and J. Gorman, "Survival's Ick Factor," *New York Times,* January 23, 2012.

"There is nothing": Curtis interview.

Indeed, Harvard psychologist: V. Curtis and A. Biran, "Dirt, Disgust, and Disease: Is Hygiene in Our Genes?," *Perspectives in Biology and Medicine* 44, no. 1 (Winter 2001): 22.

"Obviously life experience": Curtis interview.

157 *Like our sex drive:* Curtis, *Don't Look, Don't Touch, Don't Eat,* chapter 1.

Or if, after: Ibid., chapter 3.

By the time we reach adulthood: Ibid.

158 *Standards of cleanliness:* Ibid., chapter 1.

Experiments that tap: M. Schaller, D. R. Murray, and A. Bangerter, "Implications of the Behavioural Immune System for Social Behaviour and Human Health in the Modern World," *Philosophical Transactions of the Royal Society B* 370 (2015): 3, http://dx.doi.org/10.1098/rstb.2014.0105.

Personality adds another wrinkle: Curtis, *Don't Look, Don't Touch, Don't Eat,* chapter 3.

The easily disgusted: V. Curtis, "Why Disgust Matters," *Philosophical Transactions of the Royal Society B* 366 (2011): 3482–84, doi: 10–1098/rstb.2011.0165.

159 *If Curtis is right:* Curtis, *Don't Look, Don't Touch, Don't Eat,* chapter 3.

160 *Shining a bright light:* Curtis, "Why Disgust Matters," 3482–83.

Interestingly, women are: Curtis, *Don't Look, Don't Touch, Don't Eat,* chapter 1.

Perhaps related to: Rick Nauert, "Anxiety More Common in Women," Psych Central, http://psychcentral.com/news/2006/10/06/anxiety-more-common-in-women/312.html. See also "Mental Health Statistics: Men and Women," Mental Health Foundation, http://www.mentalhealth.org.uk/help-information/mental-health-statistics/men-women/.

Our disgustability and: Rachel Herz, *That's Disgusting: Unraveling the Mysteries of Repulsion* (New York: W. W. Norton, 2012), 504.

Just as hunger can: Ibid.; also Curtis, *Don't Look, Don't Touch, Don't Eat,* chapter 3.

161 *Similar findings have:* C. Borg and P. J. de Jong, "Feelings of Disgust and Disgust-Induced Avoidance Weaken Following Induced Sexual Arousal in Women," *PLoS One* 7 (September 2012): 1–8, e44111.

The huge disparity: "Ewwwww! UCLA Anthropologist Studies Evolution's Disgusting Side," UCLA Newsroom, March 27, 2007, http://newsroom.ucla.edu/releases/Ewwwww-UCLA-Anthropologist-Studies-7821.

How repulsive you find: Herz, *That's Disgusting,* chapter 4.

162 *Disgust can bias our perceptions:* Gary D. Sherman, "The Faintest Speck of Dirt: Disgust Enhances the Detection of Impurity," 25th American Psychological Science Society Meeting, Washington, DC, May 26, 2013. Also see G. D. Sher-

man, J. Haidt, and Gerald L. Clore, "The Faintest Speck of Dirt: Disgust Enhances the Detection of Impurity," *Psychological Science* 23, no. 12 (2012): 1513, doi: 10.1177/0956797612445318.

163   *High levels of disgust:* Curtis, *Don't Look, Don't Touch, Don't Eat,* chapter 3.

   For example, clothes on: K. J. Eskine, A. Novreske, and M. Richards, "Moral Contagion in Everyday Interpersonal Encounters," *Journal of Experimental Social Psychology* 49 (2013): 949.

   Grocery shoppers, for instance: David Pizarro, interview by the author, April 20, 2015.

   Succinctly encapsulating their views: "Food for Thought: Paul Rozin's Research and Teaching at Penn," *Penn Arts and Sciences* (Fall 1997), http://www.sas.upenn.edu/sasalum/newsltr/fall97/rozin.html.

164   *When I spoke to him:* Ibid.; also Rozin interview.

10. PARASITES AND PREJUDICE

165   *Parasites held no interest:* M. Schaller, interview by the author, February 1, 2011, and June 4, 2012.

166   *Schaller tried another tactic:* M. Faulkner et al., "Evolved Disease-Avoidance Mechanisms and Contemporary Xenophobic Attitudes," *Group Processes and Intergroup Relations* 7, no. 4 (2004): 344–45, doi: 10.1177/1368430204046142. See also M. Schaller and S. L. Neuberg, "Danger, Disease, and the Nature of Prejudice(s)," in *Advances in Experimental Social Psychology,* ed. M. Zanna and J. Olson (San Diego: Academic Press, 2012), 19–20.

167   *Drawing on more than:* M. Schaller, interview by the author, May 2008. Also see J. Faulkner and M. Schaller, "Evolved Disease-Avoidance Processes and Contemporary Anti-Social Behavior: Prejudicial Attitudes and Avoidance of People with Physical Disabilities," *Journal of Nonverbal Behavior* 27, no. 2 (Summer 2003): 65, and J. H. Park, M. Schaller, and C. S. Crandall, "Pathogen-Avoidance Mechanisms and the Stigmatization of Obese People," *Evolution and Human Behavior* 28 (2007): 410–14.

168   *In cognitive science:* J. Ackerman, interview by the author, August 8, 2012. Also see J. M. Ackerman et al., "A Pox on the Mind: Disjunction of Attention and Memory in the Processing of Physical Disfigurement," *Journal of Experimental Social Psychology* 45 (2009): 478–79.

   "They all look the same to me": Schaller and Neuberg, "Danger, Disease, and the Nature of Prejudice(s)," 14.

169   *But the unique traits:* Ackerman interview.

   To Schaller, it's "mind-boggling": M. Schaller, interview by the author, June 2012, and in Vancouver, September 10, 2013.

   While he clearly thinks: Schaller interview, May 2008.

170   *Compared to people:* M. Oaten, R. J. Stevenson, and T. I. Case, "Disgust as a Disease-Avoidance Mechanism," *Psychological Bulletin* 135, no. 2 (2009): 312.

by their own accounts: M. Schaller, D. R. Murray, and A. Bangerter, "Implications of the Behavioural Immune System for Social Behaviour and Human Health in the Modern World," *Philosophical Transactions of the Royal Society B* 370 (2015): 6, http://dx.doi.org/10.1098/rstb.2014.0105; Ackerman interview.

they more frequently: Schaller and Neuberg, "Danger, Disease, and the Nature of Prejudice(s)," 18.

and they report: Oaten, Stevenson, and Case, "Disgust as a Disease-Avoidance Mechanism," 312.

171   *These antipathies:* Schaller and Neuberg, "Danger, Disease, and the Nature of Prejudice(s)," 18–19.

The recently ill: Ibid., 17, 19.

Further research by Fessler: Daniel Fessler, interview by the author, Los Angeles, September 12, 2013.

172   *The behavioral immune system:* C. R. Mortensen et al., "Infection Breeds Reticence: The Effects of Disease Salience on Self-Perceptions of Personality and Behavioral Avoidance Tendencies," *Psychological Science* 21, no. 3 (2010): 440–45.

Still, those who fear contagion: C. D. Navarrete and D.M.T. Fessler, "Disease Avoidance and Ethnocentrism: The Effects of Disease Vulnerability and Disgust Sensitivity on Intergroup Attitudes," *Evolution and Human Behavior* 27 (2006): 272.

Intriguingly, several studies: Ackerman interview; see J. Y. Huang et al., "Immunizing Against Prejudice: Effects of Disease Protection on Attitudes Toward Out-Groups," *Psychological Science* 22, no. 12 (2011): 1550–56.

173   *Political scientists are:* Michael Bang Petersen and Lene Aarøe, interview by the author, Miami Beach, Florida, July 19, 2013.

175   *Schaller favors that:* Schaller interview, May 2008.

The world blamed: Brian Alexander, "Amid Swine Flu Outbreak, Racism Goes Viral," MSNBC.com, last modified May 1, 2009, http://www.nbcnews.com/id/30467300/ns/health-cold_and_flu/t/amid-swine-flu-outbreak-racism-goes-viral/#.U98FOkjY3RB; Donald G. McNeil Jr., "Finding a Scapegoat When Epidemics Strike," *New York Times,* September 1, 2009.

And in 2014: Lindsey Boerma, "Republican Congressman: Immigrant Children Might Carry Ebola," CBS News, last modified August 5, 2014, http://www.cbsnews.com/news/republican-congressman-immigrant-children-might-carry-ebola/; Maggie Fox, "Vectors or Victims? Docs Slam Rumors That Migrants Carry Disease," MSNBC News, last modified July 9, 2014, http://www.nbcnews.com/storyline/immigration-border-crisis/vectors-or-victims-docs-slam-rumors-migrants-carry-disease-n152216.

176 *The ancient Romans:* Schaller and Neuberg, "Danger, Disease, and the Nature of Prejudice(s)," 19.

Jews — history's favorite scapegoats: "Films, Nazi Antisemitic," Yad Vashem Organization, http://www.yadvashem.org/odot_pdf/Microsoft%20Word%20-%205850.pdf.

in the United States, law-abiding: A More Perfect Union, an exhibition on the Japanese American internment in World War II that toured the U.S. in the 1980s, sponsored by the Rockefeller Foundation, AT&T Foundation, and the Smithsonian Institution; http://amhistory.si.edu/perfectunion/resources /touring.html.

In 1994, Rwanda: Rachel Herz, *That's Disgusting: Unraveling the Mysteries of Repulsion* (New York: W. W. Norton, 2012), 112.

Indeed, some of the most bitterly: Brit Bennett, "Who Gets to Go to the Pool?," *New York Times,* June 10, 2015; see also Vio Celaya, *First Mexican* (Lincoln, NE: iUniverse, 2005), 4.

177 *Gerald L. Clore and:* W. Herbert, "The Color of Sin — Why the Good Guys Wear White," *Scientific American* (November 1, 2009), http://www.scientific american.com/article.cfm?id=the-color-of-sin.

To complete this: Huang et al., "Immunizing Against Prejudice," 1555.

Cancer patients often: K. McAuliffe, "The Breast Cancer Generation," *More,* September 1997.

178 *The sense of shame:* Valerie Curtis, *Don't Look, Don't Touch, Don't Eat: The Science Behind Revulsion* (Chicago: University of Chicago Press, 2013), Kindle edition, chapter 6.

Even those who care: Valerie Curtis, interview by the author, July 1, 2013.

Encouragingly, research suggests: Oaten, Stevenson, and Case, "Disgust as a Disease-Avoidance Mechanism," 308.

179 *By enlisting the:* M. Schaller, interview by the author, September 10, 2010; M. Schaller et al., "Mere Visual Perception of Other People's Disease Symptoms Facilitates a More Aggressive Immune Response," *Psychological Science* 21, no. 5 (2010): 649–52.

In an Australian investigation: R. J. Stevenson et al., "The Effect of Disgust on Oral Immune Function," *Psychophysiology* 48 (2011): 900–907.

180 *A similar British study:* Herz, *That's Disgusting,* 133.

If, as this research: Schaller interview, September 10, 2010.

11. PARASITES AND PIETY

181 *The young man was:* David Pizarro, interview by the author, April 20, 2015.

182 *Sometimes we feel:* Jonathan Haidt, *The Righteous Mind: Why Good People Are*

Divided by Politics and Religion (New York: Pantheon, 2012), Kindle edition, chapter 2.

Pizarro has a deep: Pizarro interview.

183 *Values actually:* G. Miller, "The Roots of Morality," *Science* 320 (May 9, 2008): 734.

Premarital sex: T. G. Adams, P. A. Stewart, and J. C. Blanchar, "Disgust and the Politics of Sex: Exposure to a Disgusting Odorant Increases Politically Conservative Views on Sex and Decreases Support for Gay Marriage," *PLoS One* 9, no. 5 (2014): e95572, doi:10.1371/journal.pone.0095572. Also see Haidt, *The Righteous Mind,* and Y. Inbar and D. Pizarro, "Pollution and Purity in Moral and Political Judgment," in *Advances in Experimental Moral Psychology: Affect, Character, and Commitments,* ed. J. Wright and H. Sarkissian (London: Continuum, 2014), 121.

In a trial conducted: Haidt, *The Righteous Mind,* chapter 3.

184 *When subjects in one:* Adams, Stewart, and Blanchar, "Disgust and the Politics of Sex."

There's a clear pattern: M. Schaller, D. R. Murray, and A. Bangerter, "Implications of the Behavioural Immune System for Social Behaviour and Human Health in the Modern World," *Philosophical Transactions of the Royal Society B* 370 (2015): 4, http://dx.doi.org/10.1098/rstb.2014.0105.

Disease cues may even: Adams, Stewart, and Blanchar, "Disgust and the Politics of Sex."

In light of these: E. G. Helzer and D. A. Pizarro, "Dirty Liberals! Reminders of Physical Cleanliness Influence Moral and Political Attitudes," *Psychological Science* 22, no. 4 (2011): 517.

185 *Intriguing as these results:* Pizarro interview.

186 *The highly disgustable:* Y. Inbar, D. A. Pizarro, and Paul Bloom, "Conservatives Are More Easily Disgusted Than Liberals," *Cognition and Emotion* 23, no. 4 (2009): 720, http://dx.doi.org/10.1080/02699930802110007. Also see Y. Inbar et al., "Disgust Sensitivity, Political Conservatism and Voting," *Social Psychological and Personality Science* 5 (2012): 537–44, and D. R. Murray and M. Schaller, "Threat(s) and Conformity Deconstructed: Perceived Threat of Infectious Disease and Its Implications for Conformist Attitudes and Behavior," *European Journal of Social Psychology* 42 (2012): 181, doi: 10.1002/ejsp.863.

When shown pictures: Kevin B. Smith et al., "Disgust Sensitivity and the Neurophysiology of Left-Right Political Orientations," *PLoS One* 6, no. 10 (October 2011): e25552. Also see Nicholas Kristof, "Our Politics May Be All in Our Head," *New York Times,* February 13, 2010.

Compared to liberals: Douglas R. Oxley et al., "Political Attitudes Vary with Physiological Traits," *Science* 321, no. 19 (September 19, 2008): 1667–70.

conservatives typically view: Haidt, *The Righteous Mind,* chapter 12.

187   *In a study of 237 Dutch:* C. J. Brenner and Y. Inbar, "Disgust Sensitivity Predicts Political Ideology and Policy Attitudes in the Netherlands," *European Journal of Social Psychology* 45 (2015): 27–38, doi: 10.1002/ejsp.2072.
      Conducted by a team: Y. Inbar et al., "Disgust Sensitivity, Political Conservatism and Voting," 542.

188   *A noteworthy example:* Peter Liberman and David Pizarro, "All Politics Is Olfactory," *New York Times,* October 23, 2010.
      To Gary D. Sherman: Gary D. Sherman and Gerald L. Clore, "The Color of Sin: White and Black Are Perceptual Symbols of Moral Purity and Pollution," *Psychological Science* 20, no. 8 (2009): 1019–25. Also see W. Herbert, "The Color of Sin — Why the Good Guys Wear White," *Scientific American* (November 1, 2009).

190   *"The darkness-contamination-evil":* Gerald L. Clore, interview by the author, December 30, 2015.
      Visceral disgust — that: Rachel Herz, *That's Disgusting: Unraveling the Mysteries of Repulsion* (New York: W. W. Norton, 2012), 63–65.
      In one notable: C. T. Dawes et al., "Neural Basis of Egalitarian Behavior," *Proceedings of the National Academy of Sciences* 109, no. 17 (April 24, 2012): 6479–83, doi: 10.1073/pnas.1118653109. Also see Roger Highfield, "The Robin Hood Impulse," *Telegraph,* April 11, 2007.
      The anterior insula: Valerie Curtis, *Don't Look, Don't Touch, Don't Eat: The Science Behind Revulsion* (Chicago: University of Chicago Press, 2013), Kindle edition, chapter 5.

191   *In addition, it's:* F. Kodaka et al., "Effect of Cooperation Level of Group on Punishment for Non-Cooperators: A Functional Magnetic Resonance Imaging Study," *PLoS One* 7, no. 7 (July 2012): e41338, doi: 10.1371/journal.pone.0041338.
      These kinds of: Sandra Blakeslee, "A Small Part of the Brain, and Its Profound Effects," *New York Times,* February 6, 2007.
      Some scientists think: J. S. Borg, D. Lieberman, and K. A. Kiehl, "Infection, Incest, and Iniquity: Investigating the Neural Correlates of Disgust and Morality," *Journal of Cognitive Neuroscience* 20, no. 9 (2008): 1529–46.
      Psychopaths — whose ranks: K. A. Kiehl and J. W. Buckholtz, "Inside the Mind of a Psychopath," *Scientific American Mind* 21, no. 4 (September/October 2010): 22–29. Also Kent A. Kiehl, interview by the author, August 12, 2015.
      People with Huntington's: Kiehl interview; Herz, *That's Disgusting,* chapter 3.
      Interestingly, women rarely: Ibid.

192   *Evidence from prehistoric:* Curtis, *Don't Look, Don't Touch, Don't Eat,* chapter 4.
      From this point: Valerie Curtis, interview by the author, July 1, 2013.

193   *Explaining why we:* Haidt, *The Righteous Mind,* chapter 9; also see Curtis, *Don't Look, Don't Touch, Don't Eat,* chapter 5.
      "If you're greedy": Curtis interview.

Darwin thought our: Haidt, *The Righteous Mind,* chapter 9.

194 *Disgust's use to curb:* Curtis interview.

195 *About ten thousand years ago:* Jonathan Hawks, interview by the author, Madison, Wisconsin, February 12, 2008.

As these risks mounted: F. Thomas et al., "Can We Understand Modern Humans Without Considering Pathogens?," *Evolutionary Applications* 5, no. 4 (June 2012): 368–79.

196 *Consider the fate:* Jared Diamond, *Guns, Germs, and Steel* (New York: W. W. Norton, 1997), 210.

One of the oldest: John Durant, *The Paleo Manifesto: Ancient Wisdom for Lifelong Health* (New York: Harmony, 2013), Kindle edition, chapter 4.

198 *But Hinduism, which:* Haidt, *The Righteous Mind,* chapter 5.

199 *One of the most surprising:* K. McAuliffe, "Are We Still Evolving?," *Discover,* March 2009.

200 *Leon Kass, chairman:* Leon Kass, "The Wisdom of Repugnance," *New Republic,* June 2, 1997.

201 *Orthodox Jews:* Herz, *That's Disgusting,* 171–72.

From a legal standpoint: Pizarro interview.

202 *Even more troublesome:* E. J. Horberg et al., "Disgust and the Moralization of Purity," *Journal of Personality and Social Psychology* 97, no. 6 (2009): 965.

A related study: L. van Dillen and G. Vanderveen, "Moral Integrity and Emotional Vigilance," paper presented at the biannual conference of the International Society for Research on Emotion, July 8–10, 2015, http://www.isre2015 .org/sites/default/files/van%20Dillen.pdf.

"I've been approached: Pizarro interview.

It may take long: C. Helion and D. Pizarro, "Beyond Dual-Processes: The Interplay of Reason and Emotion in Moral Judgment," in *Springer Handbook for Neuroethics,* ed. Jens Clausen and Neil Levy (New York: Springer Reference, 2015), 113.

12. THE GEOGRAPHY OF THOUGHT

205 *People living in:* Randy Thornhill, interview by the author, fall 2008.

206 *That team was headed:* Mark Schaller, interview by the author, September 10, 2013.

By superimposing this: M. Schaller and D. R. Murray, "Pathogens, Personality, and Culture: Disease Prevalence Predicts Worldwide Variability in Sociosexuality, Extraversion, and Openness to Experience," *Journal of Personality and Social Psychology* 95, no. 1 (July 2008): 212–21, doi: 10.1037/0022-3514.95.1.212.

207 *Their traits, Schaller:* Schaller interview.

208 *The results suggested:* C. L. Fincher et al., "Pathogen Prevalence Predicts Human Cross-Cultural Variability in Individualism/Collectivism," *Proceedings of the Royal Society B* 275 (2008): 1279–85, doi:10.1098/rspb.2008.0094.

"*We're not saying*": C. Fincher, interview by the author, 2008.

Thornhill, who grew: Thornhill interview.

209 *Cornell psychologist David Pizarro:* David Pizarro, interview by the author, April 20, 2015.

Steven Pinker: Steven Pinker, interview by the author, July 19, 2013, and November 3, 2015.

Of course, it has: Valerie Curtis, interview by the author, July 1, 2013.

The scientists have: C. L. Fincher and R. Thornhill, "Parasite-Stress Promotes In-Group Assortative Sociality: The Cases of Strong Family Ties and Heightened Religiosity," *Behavioral and Brain Sciences* 35 (2012): 62, 72–74, doi: 10.1017/S0140525X11000021.

210 "*Religious scholars*": Thornhill interview.

211 *To test their:* R. Thornhill, C. L. Fincher, and D. Aran, "Parasites, Democratization, and the Liberalization of Values Across Contemporary Countries," *Biological Reviews* 84 (2009): 113–15.

If we are to believe: K. Letendre, C. L. Fincher, and R. Thornhill, "Does Infectious Disease Cause Global Variation in the Frequency of Intrastate Armed Conflict and Civil War?," *Biological Reviews* 85 (2010): 669–83; and R. Thornhill and C. L. Fincher, "Parasite Stress Promotes Homicide and Child Maltreatment," *Philosophical Transactions of the Royal Society B* 366 (2011): 3466–77, doi: 10.1098/rstb.2011.0052.

"*Who would have*": Thornhill interview.

212 *That has occasionally:* Randy Thornhill, interview by the author, Miami Beach, Florida, July 20, 2013.

Schaller and Murray have: Mark Schaller, interview by the author, October 30, 2012.

Instead of doing: D. R. Murray, M. Schaller, and P. Suedfeld, "Pathogens and Politics: Further Evidence That Parasite Prevalence Predicts Authoritarianism," *PLoS One* 8, no. 5 (May 2013): e62275.

213 *They also conducted:* D. R. Murray, R. Trudeau, and M. Schaller, "On the Origins of Cultural Differences in Conformity: Four Tests of the Pathogen Prevalence Hypothesis," *Personality and Social Psychology Bulletin* 37, no. 3 (2011): 318–29, doi: 10.1177/0146167210394451.

"*Value systems*": Mark Schaller, interview by the author, October 2010.

Although there's: Schaller interview, October 30, 2012.

214 *How parasite stress gets:* Schaller interview, October 2010.

Fincher speculates: Fincher interview.

Thornhill suspects: Randy Thornhill, interview by the author, August 11, 2008.

215 *A smattering of:* Charles Nunn, interview by the author, April 15, 2015; R. H.

Griffin and C. L. Nunn, "Community Structure and the Spread of Infectious Disease in Primate Social Networks," *Evolutionary Ecology* 26 (2012): 779–800. *Evolutionary psychologist:* Daniel Fessler, interview by the author, Los Angeles, September 12, 2013. Valerie Curtis, *Don't Look, Don't Touch, Don't Eat: The Science Behind Revulsion* (Chicago: University of Chicago Press, 2013), Kindle edition, chapter 2.

For example, anthropologist: Fincher and Thornhill, "Parasite-Stress Promotes In-Group Assortative Sociality."

216   *Thornhill, for his:* Thornhill interview, July 20, 2013.

217   *A group of ethicists:* Russell Powell, Steve Clarke, and Julian Savulescu, "An Ethical and Prudential Argument for Prioritizing the Reduction of Parasite-Stress in the Allocation of Health Care Resources," *Behavioral and Brain Sciences* 35 (2012): 90–91, doi: 10.1017/S0140525X11001026.

INDEX